MODERN
CONTROL
SYSTEMS

MODERN CONTROL SYSTEMS

An Introduction

By
S. M. Tripathi

JONES & BARTLETT
LEARNING

©2008 by INFINITY SCIENCE PRESS LLC
An imprint of Jones & Bartlett Learning, LLC

Jones & Bartlett Learning, LLC
40 Tall Pine Drive
Sudbury, MA 01776
Tel. 978-443-5000
info@jblearning.com
www.jblearning.com

Modern Control Systems by S.M.Tripathi
ISBN: 978-1-934015-21-6

Library of Congress Cataloging-in-Publication Data
Tripathi, S. (Saurabh Mani), 1984-
  Modern control systems / S. Tripathi.
    p. cm.
  ISBN 978-1-934015-21-6 (hardcover)
 1. Automatic control.   2. Control theory.   I. Title.
  TJ213.T66 2008
  629.8-—dc22

2008012204

8 9 0 1 2 3 4

Dedicated to
Dr. Saurabh Basu (My Teacher)
Smt. Shashi Tripathi & Shri Lal Bihari Tripathi (My Parents)
Kaustubh Mani Tripathi (My Brother)

CONTENTS

STATE-VARIABLE ANALYSIS OF CONTINUOUS-TIME SYSTEMS

Chapter 1

1.1   INTRODUCTION

The methods of analysis and the design of feedback control systems; such as root locus, and Bode and Nyquist plots, require the physical system to be modelled in the form of a transfer function. Although the transfer function approach of analysis is a simple and powerful technique, it suffers from certain drawbacks:

(a) The transfer function model is only applicable to linear time-invariant systems.

(b) A transfer function is only defined under zero initial conditions; meaning that the system is initially **at rest** and the time solution obtained is, in a general form, due to input only. However, for multi-input-multi-output, the systems are initially not at rest. Hence, as a conclusion, the transfer function model is generally restricted to single-input-single-output (SISO) systems.

(c) A transfer function technique only reveals the system output for a given input and does not provide any information regarding the **internal state** of the system.

Thus, the classical design methods (root locus and frequency domain methods) based on the transfer function approach are **inadequate** and not convenient.

The limitations listed above made us feel the need for a more general mathematical representation of a control system, which, along with the

1

output, yields information about the state of the system–variables at some predetermined points along the flow of signals. This leads to the development of the **state-variable approach**, which has the following advantages over the **classical approach:**

(a) It is a direct **time-domain approach**. Thus, this approach is suitable for digital computer computations.
(b) It is a very powerful technique for the design and analysis of linear or nonlinear, time-variant or time-invariant, and **SISO** or **MIMO** systems.
(c) In this technique, the nth order differential equations can be expressed as 'n' equations of first order. Thus, making the solutions easier.
(d) Using this approach, the system can be designed for optimal conditions with respect to given **performance indices**.

In spite of the above discussions, the state-variable approach cannot completely replace the classical approach.

1.2   DEFINITIONS CONCERNING THE STATE-SPACE APPROACH

1.2.1   State

The **state** of a dynamical system is a minimal set of variables $x_1(t)$, $x_2(t)$, $x_3(t)$ ...... $x_n(t)$ such that the knowledge of these variables at $t = t_0$ (initial condition), together with the knowledge of inputs $u_1(t), u_2(t), u_3(t)$ ...... $u_m(t)$ for $t \geq t_0$, completely determines the behavior of the system for $t > t_0$.

1.2.2   State-Variables

The variables $x_1(t), x_2(t), x_3(t)$ ...... $x_n(t)$ such that the knowledge of these variables at $t = t_0$ (initial condition), together with the knowledge of inputs $u_1(t), u_2(t), u_3(t)$ ...... $u_m(t)$ for $t \geq t_0$, completely determines the behavior of the system for $t > t_0$; are called **state-variables**. In other words, the variables that determine the **state** of a dynamical system, are called **state-variables**.

1.2.3   State-Vector

If n state variables $x_1(t)$, $x_2(t)$, $x_3(t)$ ...... $x_n(t)$ are necessary to determine the behavior of a dynamical system, then these n state-variables can be considered as n components of a vector $\mathbf{x}(t)$, called **state-vector**.

1.2.4 State-Space

The n dimensional space, whose elements are the n state-variables, is called **state-space**. Any state can be represented by a **point** in the state-space.

1.3 BLOCK-DIAGRAM REPRESENTATION OF A GENERAL CONTROL SYSTEM

In state-variable formulation of a system, the state-variables are usually represented by $x_1(t)$, $x_2(t)$, $x_3(t)$ ......; the inputs by $u_1(t)$, $u_2(t)$, $u_3(t)$, ...; and the outputs by $y_1(t)$, $y_2(t)$ ...... The state-space representation of a system that has m inputs, p outputs, and n-state variables may be visualized in block-diagram form as shown in Figure 1.1. For notational economy, the different variables may be represented by the input vector $\mathbf{u}(t)$, output vector $\mathbf{y}(t)$, and the state vector $\mathbf{x}(t)$, where

$$\mathbf{u}(t) = \begin{bmatrix} u_1(t) \\ u_2(t) \\ \vdots \\ u_m(t) \end{bmatrix}_{m \times 1} \; ; \; \mathbf{y}(t) = \begin{bmatrix} y_1(t) \\ y_2(t) \\ \vdots \\ y_p(t) \end{bmatrix}_{p \times 1} \; ; \; \mathbf{x}(t) = \begin{bmatrix} x_1(t) \\ x_2(t) \\ \vdots \\ x_n(t) \end{bmatrix}_{n \times 1} .$$

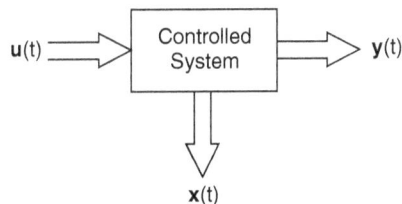

FIGURE 1.1 Block-diagram representation of a general control system.

1.4 STATE MODEL OF A GENERAL CONTROL SYSTEM

For a general system of Figure 1.1, the state representation can be arranged in the form of n first-order differential equations as

$$\left.\begin{array}{l}\dot{x}_1(t) = f_1(x_1(t), x_2(t), \ldots\ldots x_n(t)\,;\, u_1(t), u_2(t), \ldots\ldots u_m(t)) \\[4pt] \dot{x}_2(t) = f_2(x_1(t), x_2(t), \ldots\ldots x_n(t)\,;\, u_1(t), u_2(t), \ldots\ldots u_m(t)) \\[4pt] \quad\vdots \qquad\qquad \vdots \qquad\qquad\qquad \vdots \\[4pt] \dot{x}_n(t) = f_n(x_1(t), x_2(t), \ldots\ldots x_n(t)\,;\, u_1(t), u_2(t), \ldots\ldots u_m(t)) \end{array}\right\}. \qquad (1.1)$$

Integrating (1.1), we have

$$\left. x_i(t) = x_i(t_0) + \int_{t_0}^{t} f_i(x_1(\tau), x_2(\tau), \ldots\ldots x_n(\tau)\,;\, u_1(\tau), u_2(\tau), \ldots\ldots u_m(\tau))d\tau \right\};$$

$$i = 1, 2, 3, . \, n.$$

Thus, the n state-variables and, the state of the system can uniquely be determined at any $t > t_0$, provided each state-variable is known at $t = t_0$ and all the m control forces are known throughout the interval t_0 to t.

The n differential equations of (1.1) may be written in vector form as

$$\dot{\mathbf{x}}(t) = \mathbf{f}(\mathbf{x}(t), \mathbf{u}(t)) \qquad (1.2)$$

where $\quad \mathbf{x}(t) = \begin{bmatrix} x_1(t) \\ x_2(t) \\ \vdots \\ x_n(t) \end{bmatrix}_{n\times 1} \quad ;\ \text{State Vector}$

$$\mathbf{u}(t) = \begin{bmatrix} u_1(t) \\ u_2(t) \\ \vdots \\ u_m(t) \end{bmatrix}_{m\times 1} \quad ;\ \text{Input Vector}$$

and $\quad \mathbf{f}(.) = \begin{bmatrix} \mathbf{f}_1(.) \\ \mathbf{f}_2(.) \\ \vdots \\ \mathbf{f}_n(.) \end{bmatrix}_{n\times 1} \quad ;\ \text{Function Vector.}$

Equation (1.2) is the **state equation** for **time-invariant systems**. However, for **time-varying systems**, the function vector **f**(.) is dependent on time as well, and the vector equation may be given as

$$\dot{\mathbf{x}}(t) = \mathbf{f}(\mathbf{x}(t), \mathbf{u}(t), t). \tag{1.3}$$

Equation (1.3) is the **state equation** for **time-varying systems**.

The output **y**(t) can, in general, be expressed in terms of the state vector **x**(t) and input vector **u**(t) as:

For **time-invariant systems**:

$$\mathbf{y}(t) = \mathbf{g}(\mathbf{x}(t), \mathbf{u}(t)) \tag{1.4}$$

For **time-varying systems**:

$$\mathbf{y}(t) = \mathbf{g}(\mathbf{x}(t), \mathbf{u}(t), t) \tag{1.5}$$

where $\mathbf{y}(t) = \begin{bmatrix} y_1(t) \\ y_2(t) \\ \vdots \\ y_p(t) \end{bmatrix}_{p \times 1}$; Output Vector

$\mathbf{u}(t) = \begin{bmatrix} u_1(t) \\ u_2(t) \\ \vdots \\ u_m(t) \end{bmatrix}_{m \times 1}$; Input Vector

and $\mathbf{g}(\cdot) = \begin{bmatrix} g_1(\cdot) \\ g_2(\cdot) \\ \vdots \\ g_p(\cdot) \end{bmatrix}_{p \times 1}$; Function Vector.

Equations (1.4) and (1.5) are the **output equations** for time-invariant and time-varying systems, respectively.

The state equations and output equations together constitute the **state model** of the system. Thus, the state model of a general control system (shown in Figure 1.1) is given by the following equations:

For **time-invariant systems:**

$$\dot{\mathbf{x}}(t) = \mathbf{f}(\mathbf{x}(t), \mathbf{u}(t)); \qquad \text{State Equation} \qquad (1.6\ A)$$

$$\mathbf{y}(t) = \mathbf{g}(\mathbf{x}(t), \mathbf{u}(t)); \qquad \text{Output Equation} \qquad (1.6\ B)$$

For **time-varying systems:**

$$\dot{\mathbf{x}}(t) = \mathbf{f}(\mathbf{x}(t), \mathbf{u}(t), t) \qquad \text{State Equation} \qquad (1.7\ A)$$

$$\mathbf{y}(t) = \mathbf{g}(\mathbf{x}(t), \mathbf{u}(t), t) \qquad \text{Output Equation} \qquad (1.7\ B)$$

1.5 STATE MODEL OF A LINEAR MULTI-INPUT-MULTI-OUTPUT SYSTEM

The state model of a linear, time-invariant MIMO system is a special case of the general time-invariant state model of Equations (1.6). In state representation, the derivative of each state variable can be written as a linear combination of system states and inputs i.e.,

$$\left. \begin{array}{l} \dot{x}_1(t) = a_{11}x_1(t) + a_{12}x_2(t) + \ldots\ldots + a_{1n}x_n(t) + b_{11}u_1(t) + b_{12}u_2(t) + \ldots\ldots + b_{1m}u_m(t) \\ \dot{x}_2(t) = a_{21}x_1(t) + a_{22}x_2(t) + \ldots\ldots + a_{2n}x_n(t) + b_{21}u_1(t) + b_{22}u_2(t) + \ldots\ldots + b_{2m}u_m(t) \\ \quad\vdots \qquad\qquad \vdots \qquad\qquad\qquad\qquad \vdots \\ \dot{x}_n(t) = a_{n1}x_1(t) + a_{n2}x_2(t) + \ldots\ldots + a_{nn}x_n(t) + b_{n1}u_1(t) + b_{n2}u_2(t) + \ldots\ldots + b_{nm}u_m(t) \end{array} \right\}$$

$$(1.8)$$

where coefficients a_{ij}; $i = 1, 2, \ldots\ldots, n$; $j = 1, 2, \ldots\ldots, n$ and b_{ik}; $i = 1, 2, \ldots\ldots n$; $k = 1, 2, \ldots\ldots, m$ are constants. Equations (1.8) may be written in vector-matrix form as

$$\dot{\mathbf{x}}(t) = \mathbf{A}\mathbf{x}(t) + \mathbf{B}\mathbf{u}(t); \qquad\qquad \text{State Equation}$$

where $\quad \mathbf{x}(t) = \begin{bmatrix} x_1(t) \\ x_2(t) \\ \vdots \\ x_n(t) \end{bmatrix}_{n \times 1}$; State Vector

$\mathbf{u}(t) = \begin{bmatrix} u_1(t) \\ u_2(t) \\ \vdots \\ u_m(t) \end{bmatrix}_{m \times 1}$; Input Vector

$\mathbf{A} = \begin{bmatrix} a_{11} & a_{12} & \cdots\cdots & a_{1n} \\ a_{21} & a_{22} & \cdots\cdots & a_{2n} \\ \vdots & \vdots & \cdots\cdots & \vdots \\ a_{n1} & a_{n2} & \cdots\cdots & a_{nn} \end{bmatrix}_{n \times n}$; System Matrix

and $\quad \mathbf{B} = \begin{bmatrix} b_{11} & b_{12} & \cdots\cdots & b_{1m} \\ b_{21} & b_{22} & \cdots\cdots & b_{2m} \\ \vdots & \vdots & \cdots\cdots & \vdots \\ b_{n1} & b_{n2} & \cdots\cdots & b_{nm} \end{bmatrix}_{n \times m}$; Input Matrix.

Similarly, the output variables can be written as a linear combination of system states and inputs, i.e.,

$$y_1(t) = c_{11}x_1(t) + c_{12}x_2(t) + \cdots\cdots + c_{1n}x_n(t) + d_{11}u_1(t) + d_{12}u_2(t) + \cdots\cdots + d_{1m}u_m(t)$$
$$y_2(t) = c_{21}x_1(t) + c_{22}x_2(t) + \cdots\cdots + c_{2n}x_n(t) + d_{21}u_1(t) + d_{22}u_2(t) + \cdots\cdots + d_{2m}u_m(t)$$
$$\vdots \qquad\qquad \vdots \qquad\qquad\qquad\qquad \vdots$$
$$y_p(t) = c_{p1}x_1(t) + c_{p2}x_2(t) + \cdots\cdots + c_{pn}x_n(t) + d_{p1}u_1(t) + d_{p2}u_2(t) + \cdots\cdots + d_{pm}u_m(t)$$

$$(1.9)$$

where, coefficients $c_{ij}; i = 1, 2, \ldots\ldots p; j = 1, 2, \ldots n$ and $d_{ik}; i = 1, 2, \ldots\ldots p; k = 1, 2, \ldots\ldots m$ are constants. Equations (1.9) may be written in vector-matrix form as

$$\mathbf{y}(t) = \mathbf{C}\mathbf{x}(t) + \mathbf{D}\mathbf{u}(t); \qquad \text{Output Equation}$$

where $\quad \mathbf{y}(t) = \begin{bmatrix} y_1(t) \\ y_2(t) \\ \vdots \\ y_p(t) \end{bmatrix}_{p \times 1}$;   Output Vector

$\mathbf{u}(t) = \begin{bmatrix} u_1(t) \\ u_2(t) \\ \vdots \\ u_m(t) \end{bmatrix}_{m \times 1}$;   Input Vector

$\mathbf{C} = \begin{bmatrix} c_{11} & c_{12} & \cdots\cdots & c_{1n} \\ c_{21} & c_{22} & \cdots\cdots & c_{2n} \\ \vdots & \vdots & \cdots\cdots & \vdots \\ c_{p1} & c_{p2} & \cdots\cdots & c_{pn} \end{bmatrix}_{p \times n}$;   Output Matrix

and $\qquad \mathbf{D} = \begin{bmatrix} d_{11} & d_{12} & \cdots\cdots & d_{1m} \\ d_{21} & d_{22} & \cdots\cdots & d_{2m} \\ \vdots & \vdots & \cdots\cdots & \vdots \\ d_{p1} & d_{p2} & \cdots\cdots & d_{pm} \end{bmatrix}_{p \times m}$;   Transmission Matrix.

The state equation and output equation together constitute the state model of the system. Thus, the state model of a linear time-invariant MIMO system is given as

$$\dot{\mathbf{x}}(t) = \mathbf{A}\mathbf{x}(t) + \mathbf{B}\mathbf{u}(t); \qquad \text{State Equation} \qquad (1.10\ A)$$

$$\mathbf{y}(t) = \mathbf{C}\mathbf{x}(t) + \mathbf{D}\mathbf{u}(t); \qquad \text{Output Equation} \qquad (1.10\ B)$$

The state model of a linear, time-varying MIMO system is of the same form as given in Equations (1.10) except for the fact that the coefficients of the matrices $\mathbf{A}, \mathbf{B}, \mathbf{C},$ and $\mathbf{D}$ are no longer constants but are the functions of time.

Thus, the state model of a linear, time-varying MIMO system is given as

$$\dot{\mathbf{x}}(t) = \mathbf{A}(t)\mathbf{x}(t) + \mathbf{B}(t)\mathbf{u}(t); \qquad \text{State Equation} \qquad (1.11\ A)$$

$$\mathbf{y}(t) = \mathbf{C}(t)\mathbf{x}(t) + \mathbf{D}(t)\mathbf{u}(t); \qquad \text{Output Equation} \qquad (1.11\ B)$$

The block-diagram representation of the state model of a linear multi-input-multi-output system is shown in Figure 1.2.

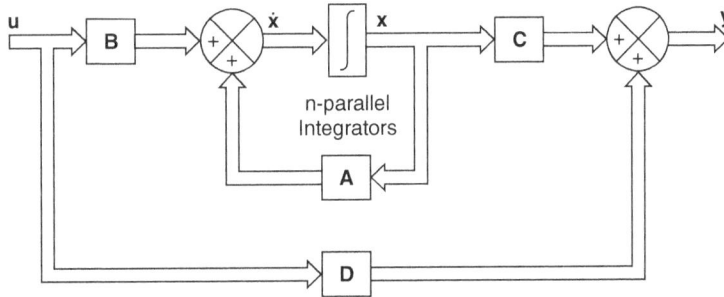

FIGURE 1.2 Block-diagram representation of the state model of a linear
multi-input-multi-output system.

1.6 STATE MODEL OF A LINEAR SINGLE-INPUT-SINGLE-OUTPUT SYSTEM

The state model of a linear single-input-single-output system can be obtained by putting $m = 1$ and $p = 1$ in the state model of a linear multi-input-multi-output system as

$$\dot{\mathbf{x}}(t) = \mathbf{A}\mathbf{x}(t) + \mathbf{B}u(t); \qquad \text{State Equation} \qquad (1.12\ A)$$

$$y(t) = \mathbf{C}\mathbf{x}(t) + du(t); \qquad \text{Output Equation} \qquad (1.12\ B)$$

where $\quad \mathbf{x}(t) = \begin{bmatrix} x_1(t) \\ x_2(t) \\ \vdots \\ x_n(t) \end{bmatrix}_{n \times 1}$; State Vector $\quad \mathbf{A} = \begin{bmatrix} a_{11} & a_{12} & \cdots\cdots & a_{1n} \\ a_{21} & a_{22} & \cdots\cdots & a_{2n} \\ \vdots & \vdots & \cdots\cdots & \vdots \\ a_{n1} & a_{n2} & \cdots\cdots & a_{nn} \end{bmatrix}_{n \times n}$;

System Matrix

$$\mathbf{B} = \begin{bmatrix} b_1 \\ b_2 \\ \vdots \\ b_n \end{bmatrix}_{n \times 1} ; \quad \text{Input Matrix} \quad \mathbf{C} = \begin{bmatrix} c_1 \ c_2 \ c_3 \ \ldots\ldots\ c_n \end{bmatrix}_{1 \times n} ; \quad \text{Output Matrix}$$

$$d = \text{Transmission Constant}$$
$$u(t) = \text{Input or Control Variable (scalar)}$$
and, $$\quad y(t) = \text{Output Variable (scalar)}.$$

The block-diagram representation of the state model of linear single-input-single-output system is shown in Figure 1.3.

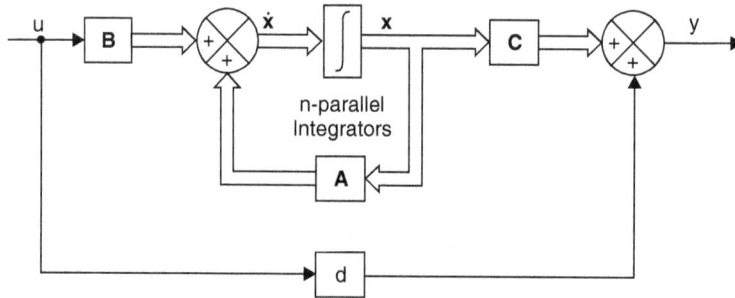

FIGURE 1.3 Block-diagram representation of the state model of the linear single-input-single-output system.

1.7 LINEARIZATION OF THE STATE EQUATION OF A GENERAL TIME-INVARIANT CONTROL SYSTEM

The state equation of a general time-invariant system is given by

$$\dot{\mathbf{x}}(t) = \mathbf{f}(\mathbf{x}(t), \mathbf{u}(t)). \tag{1.13}$$

For simplicity, we are just writing Equation (1.13) as

$$\dot{\mathbf{x}} = \mathbf{f}(\mathbf{x}, \mathbf{u}). \tag{1.14}$$

Let $(\mathbf{x}_0, \mathbf{u}_0)$ be an equilibrium point. This means the system is in equilibrium under the conditions $\mathbf{x}_0$ and $\mathbf{u}_0$, i.e.,

$$\dot{\mathbf{x}} = \mathbf{f}(\mathbf{x}_0, \mathbf{u}_0) = \mathbf{0}. \tag{1.15}$$

Equation (1.15) shows that the derivatives of all the state-variables are zero at the equilibrium point. Thus, the system continues to lie at the equilibrium point unless otherwise disturbed.

The state-equation (1.14) of a general time-invariant system can be linearized about the equilibrium point $(\mathbf{x}_0, \mathbf{u}_0)$ by expanding it into Taylor's series and neglecting second-and higher-order terms.

Thus, the ith state equation is given by

$$\dot{x}_i = f_i(\mathbf{x}_0, \mathbf{u}_0) + \sum_{j=1}^{n} \left. \frac{\partial f_i(\mathbf{x}, \mathbf{u})}{\partial x_j} \right|_{\substack{\mathbf{x} = \mathbf{x}_0 \\ \mathbf{u} = \mathbf{u}_0}} \left(x_j - x_{j0} \right) + \sum_{k=1}^{m} \left. \frac{\partial f_i(\mathbf{x}, \mathbf{u})}{\partial u_k} \right|_{\substack{\mathbf{x} = \mathbf{x}_0 \\ \mathbf{u} = \mathbf{u}_0}} \left(u_k - u_{k0} \right)$$

$$i = 1, 2, \ldots\ldots \ n.$$

Define the variations about the equilibrium point $(\mathbf{x}_0, \mathbf{u}_0)$ as

$$\tilde{x}_j = x_j - x_{j0}, \tag{1.16 A}$$

$$\tilde{u}_k = u_k - u_{k0}. \tag{1.16 B}$$

Also, from (1.16 A) we have

$$\dot{\tilde{x}}_j = \dot{x}_j.$$

Now, for $f_i(\mathbf{x}_0, \mathbf{u}_0) = 0$; the linearized i^{th} state-equation is given by

$$\dot{\tilde{x}}_i = \sum_{j=1}^{n} \left. \frac{\partial f_i(\mathbf{x}, \mathbf{u})}{\partial x_j} \right|_{\substack{\mathbf{x} = \mathbf{x}_0 \\ \mathbf{u} = \mathbf{u}_0}} \tilde{x}_j + \sum_{k=1}^{m} \left. \frac{\partial f_i(\mathbf{x}, \mathbf{u})}{\partial u_k} \right|_{\substack{\mathbf{x} = \mathbf{x}_0 \\ \mathbf{u} = \mathbf{u}_0}} \tilde{u}_k$$

When writing the above linearized state equations in vector-matrix form

$$\dot{\tilde{\mathbf{x}}} = \mathbf{A}\tilde{\mathbf{x}} + \mathbf{B}\tilde{\mathbf{u}} \tag{1.17}$$

$$\text{where} \quad \mathbf{A} = \begin{bmatrix} \dfrac{\partial f_1}{\partial x_1} & \dfrac{\partial f_1}{\partial x_2} & \cdots\cdots & \dfrac{\partial f_1}{\partial x_n} \\[2ex] \dfrac{\partial f_2}{\partial x_1} & \dfrac{\partial f_2}{\partial x_2} & \cdots\cdots & \dfrac{\partial f_2}{\partial x_n} \\[2ex] \vdots & \vdots & & \vdots \\[2ex] \dfrac{\partial f_n}{\partial x_1} & \dfrac{\partial f_n}{\partial x_2} & \cdots\cdots & \dfrac{\partial f_n}{\partial x_n} \end{bmatrix}_{n\times n}$$

$$\text{and} \quad \mathbf{B} = \begin{bmatrix} \dfrac{\partial f_1}{\partial u_1} & \dfrac{\partial f_1}{\partial u_2} & \cdots\cdots & \dfrac{\partial f_1}{\partial u_m} \\[2ex] \dfrac{\partial f_2}{\partial u_1} & \dfrac{\partial f_2}{\partial u_2} & \cdots\cdots & \dfrac{\partial f_2}{\partial u_m} \\[2ex] \vdots & \vdots & & \vdots \\[2ex] \dfrac{\partial f_n}{\partial u_1} & \dfrac{\partial f_n}{\partial u_2} & \cdots\cdots & \dfrac{\partial f_n}{\partial u_m} \end{bmatrix}_{n\times m} .$$

The matrices $\mathbf{A}$ and $\mathbf{B}$ are called the **Jacobian Matrices** and all the partial derivatives in matrices $\mathbf{A}$ and $\mathbf{B}$ are evaluated at the equilibrium point $(\mathbf{x}_0, \mathbf{u}_0)$.

1.8   HOMOGENEOUS AND NONHOMOGENEOUS LINEAR TIME-INVARIANT SYSTEMS

The state equation of a linear time-invariant system is given by

$$\dot{\mathbf{x}}(t) = \mathbf{A}\mathbf{x}(t) + \mathbf{B}\mathbf{u}(t). \tag{1.18}$$

Case 1. If $\mathbf{A}$ is a constant matrix and $\mathbf{u}(t)$ is a **zero vector**, i.e., no control forces are applied to the system, then the Equation (1.18) represents a **homogeneous** linear time-invariant system.

Case 2. If $\mathbf{A}$ is a constant matrix and $\mathbf{u}(t)$ is **nonzero vector**, i.e., control forces are applied to the system, then Equation (1.18) represents a **nonhomogeneous** linear time-invariant system.

1.9   SOLUTION OF STATE EQUATIONS FOR LINEAR TIME-INVARIANT SYSTEMS

1.9.1   State Transition Matrix (STM)

The state equation of a linear time-invariant system is given by

$$\dot{\mathbf{x}}(t) = \mathbf{A}\mathbf{x}(t) + \mathbf{B}\mathbf{u}(t).$$

For a homogeneous (unforced) system

$$\mathbf{u}(t) = \mathbf{0}$$

we have

$$\mathbf{x}(t) = \mathbf{A}\mathbf{x}(t). \tag{1.19}$$

Take the Laplace Transform on both sides

$$s\mathbf{X}(s) - \mathbf{x}(0) = \mathbf{A}\mathbf{X}(s). \tag{1.20}$$

Equation (1.20) may also be written as

or, $s\mathbf{I}\mathbf{X}(s) - \mathbf{x}(0) = \mathbf{A}\mathbf{X}(s)$

or, $[s\mathbf{I} - \mathbf{A}]\,\mathbf{X}(s) = \mathbf{x}(0)$

$$\mathbf{X}(s) = [s\mathbf{I} - \mathbf{A}]^{-1}\mathbf{x}(0).$$

Take the inverse Laplace

$$\mathbf{x}(t) = e^{\mathbf{A}t}\mathbf{x}(0). \tag{1.21}$$

Equation (1.21) gives the **solution** of the LTI **homogeneous state equation (1.19)**. From Equation (1.21) it is observed that the initial state $\mathbf{x}(0)$ at $t = 0$, is driven to a state $\mathbf{x}(t)$ at time t. This transition in state is carried out by the matrix exponential $e^{\mathbf{A}t}$. Because of this property, $e^{\mathbf{A}t}$ is termed as the **State Transition Matrix** and is denoted by $\varnothing(t)$.

Thus,

$$\phi(t) = e^{\mathbf{A}t} = \pounds^{-1}\{[s\mathbf{I} - \mathbf{A}]^{-1}\} = \pounds^{-1}\mathbf{\Phi}(s)$$

where, $\mathbf{\Phi}(s) = [s\mathbf{I} - \mathbf{A}]^{-1}$ is called the **Resolvent Matrix**.
As $e^{\mathbf{A}t}$ represents a power series of the matrix $\mathbf{A}t$, thus,

$$\phi(t) = e^{\mathbf{A}t} = \mathbf{I} + \mathbf{A}t + \frac{\mathbf{A}^2 t^2}{2!} + \ldots..$$

1.9.1.1 Properties of the State Transition Matrix

1. $\phi(0) = \mathbf{I}$.

Proof. $\qquad\qquad\qquad\qquad\qquad\qquad \phi(t) = e^{\mathbf{A}t}$

Put $t = 0$

We have $\qquad\qquad\qquad\qquad\qquad\qquad \phi(0) = \mathbf{I}$.

2. $\phi^{-1}(t) = \phi(-t)$

Proof. $\qquad\qquad\qquad\qquad\qquad\qquad \phi(t) = e^{\mathbf{A}t}$

Postmultiply both sides by $e^{-\mathbf{A}t}$.

We have $\qquad\qquad\qquad\qquad\qquad \phi(t)e^{-\mathbf{A}t} = e^{\mathbf{A}t}.e^{-\mathbf{A}t}$

or, $\qquad\qquad\qquad\qquad\qquad\qquad \phi(t)\phi(-t) = \mathbf{I}$.

Premultiply both sides by $\phi^{-1}(t)$.

We have $\qquad \phi^{-1}(t)\phi(t)\phi(-t) = \phi^{-1}(t)$

or, $\qquad\qquad\qquad\qquad \phi(-t) = \phi^{-1}(t)$

or, $\qquad\qquad\qquad\qquad \phi^{-1}(t) = \phi(-t)$.

An interesting result from this property of STM is that Equation (1.21) can be rearranged as

$$\mathbf{x}(0) = \phi(-t)\,\mathbf{x}(t),$$

which clearly means that the transition in time can take place in either direction (**bilateral in time**).

3. $\phi(t_1 + t_2) = \phi(t_1)\,\phi(t_2) = \phi(t_2)\,\phi(t_1).$

Proof. $\phi(t_1 + t_2) = e^{\mathbf{A}(t_1 + t_2)}$

$$= e^{\mathbf{A}t_1}e^{\mathbf{A}t_2} = \phi(t_1)\,\phi(t_2)$$

Again, $\phi(t_1 + t_2) = e^{\mathbf{A}(t_1 + t_2)} = e^{\mathbf{A}(t_2 + t_1)}$

$$= e^{\mathbf{A}t_2}e^{\mathbf{A}t_1} = \phi(t_2)\,\phi(t_1)$$

4. $\left[\phi(t)\right]^n = \phi(nt);\ n = \text{integer}$

Proof. $[\phi(t)]^n = e^{\mathbf{A}t}e^{\mathbf{A}t}e^{\mathbf{A}t}\dots\dots$ upto n terms

$$= e^{\mathbf{A}nt} = \phi(nt)$$

5. $\dot\phi(t) = \mathbf{A}\phi(t)$

Proof. $\dot\phi(t) = \dfrac{d}{dt}e^{\mathbf{A}t}$

$$= \mathbf{A}e^{\mathbf{A}t} = \mathbf{A}\phi(t)$$

6. $\phi(t_2 - t_1)\phi(t_1 - t_0) = \phi(t_2 - t_0) = \phi(t_1 - t_0)\phi(t_2 - t_1)$

Proof: $\phi(t_2 - t_1)\phi(t_1 - t_0) = e^{\mathbf{A}(t_2 - t_1)}\,e^{\mathbf{A}(t_1 - t_0)} = e^{\mathbf{A}(t_2 - t_1 + t_1 - t_0)}$

$$= e^{\mathbf{A}(t_2 - t_0)} = \phi(t_2 - t_0)$$

Similarly,

$$\phi\left(t_1 - t_0\right)\phi\left(t_2 - t_1\right) = \phi\left(t_2 - t_0\right)$$

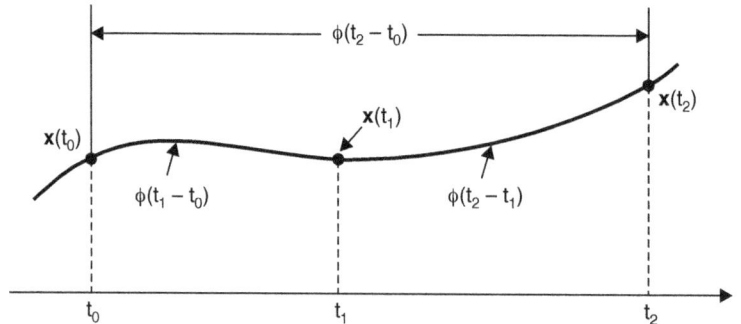

FIGURE 1.4 **Property of STM.**

This is a very important property of STM as it implies that a **state transition process** can be divided into a number of **sequential transitions**.

Figure 1.4 illustrates that the transition from $t = t_0$ to $t = t_2$ is equal to the transition from t_0 to t_1 and then from t_1 to t_2. In general, of course, the transition process can be broken up into any number of parts.

1.9.2   State Transition Equation

The state transition equation is defined as the **solution** of the linear nonhomogeneous state equation (forced system).

Consider the state equation of a linear time-invariant system

$$\dot{x}(t) = \mathbf{A}x(t) + \mathbf{B}u(t). \qquad (1.22)$$

For a nonhomogeneous (forced) system, $\mathbf{u}(t)$ is a nonzero vector, i.e., control forces are applied to the system. Thus, Equation (1.22) represents a nonhomogeneous linear time-invariant system.

The solution of Equation (1.22) can be obtained by considering two cases as follows:

Case 1: When initial state is known at $t = 0$.

Rewrite Equation (1.22) in the form

$$\dot{\mathbf{x}}(t) - \mathbf{A}\mathbf{x}(t) = \mathbf{B}u(t).$$

Multiplying both sides by $e^{-\mathbf{A}t}$, we have

$$e^{-\mathbf{A}t}\left[\dot{\mathbf{x}}(t) - \mathbf{A}\mathbf{x}(t)\right] = e^{-\mathbf{A}t}\,\mathbf{B}\mathbf{u}(t)$$

or,
$$\frac{d}{dt}[e^{-\mathbf{A}t}\,\mathbf{x}(t)] = e^{-\mathbf{A}t}\,\mathbf{B}\mathbf{u}(t).$$

Integrate both sides with respect to t between the limits 0 and t, and we have

$$e^{-\mathbf{A}t}\,\mathbf{x}(t) - \mathbf{x}(0) = \int_0^t e^{-\mathbf{A}\tau}\,\mathbf{B}\mathbf{u}(\tau)\,d\tau.$$

Premultiply both sides by $e^{\mathbf{A}t}$, and we have

$$\mathrm{x}(t) = e^{\mathbf{A}t}\mathbf{x}(0) \;+\; \int_0^t e^{\mathbf{A}(t-\tau)}\,\mathbf{B}\mathbf{u}(\tau)\,d\tau$$

or,
$$\mathrm{x}(t) = \phi(t)\mathbf{x}(0) + \int_0^t \boldsymbol{\phi}^{\mathbf{A}(t-\tau)}\,\mathbf{B}\mathbf{u}(\tau)\,d\tau. \qquad (1.23)$$

Homogeneous
Solution   Forced Solution

Case 2: When initial state is known at $t = t_0$.
Put $t = t_0$ in Equation (1.23) and solve for $\mathbf{x}(0)$.

$$\mathbf{x}(t_0) = \phi(t_0)\mathbf{x}(0) + \int_0^{t_0} \phi(t_0 - \tau)\,\mathbf{B}\mathbf{u}(\tau)\,d\tau$$

Multiplying both sides by $\phi^{-1}(t_0)$, we have

$$\phi^{-1}(t_0)\mathbf{x}(t_0) = \mathbf{x}(0) + \int_0^{t_0} \phi(-t_0)\phi(t_0 - \tau)\mathbf{B}\mathbf{u}(\tau)d\tau \qquad \because [\phi^{-1}(t_0) = \phi(-t_0)]$$

or,
$$\mathbf{x}(0) = \phi(-t_0)\mathbf{x}(t_0) - \int_0^{t_0} \phi(-t_0 + t_0 - \tau)\mathbf{B}\mathbf{u}(\tau)d\tau$$

or,
$$\mathbf{x}(0) = \phi(-t_0)\mathbf{x}(t_0) - \int_0^{t_0} \phi(-\tau)\mathbf{B}\mathbf{u}(\tau)d\tau. \qquad (1.24)$$

Put the value of $x(0)$ from Equation (1.24) in Equation (1.23), we have

$$\mathbf{x}(t) = \phi(t)\left[\phi(-t_0)\,\mathbf{x}(t_0) - \int_0^{t_0}\phi(-\tau)\,\mathbf{B}\mathbf{u}(\tau)\,d\tau\right] + \int_0^t\phi(t-\tau)\,\mathbf{B}\mathbf{u}(\tau)\,d\tau$$

or, $\quad \mathbf{x}(t) = \phi(t-t_0)\mathbf{x}(t_0) - \int_0^{t_0}\phi(t-\tau)\,\mathbf{B}\mathbf{u}(\tau)\,d\tau + \int_0^t\phi(t-\tau)\,\mathbf{B}\mathbf{u}(\tau)\,d\tau$

or, $\quad \mathbf{x}(t) = \phi(t-t_0)\mathbf{x}(t_0) + \int_{t_0}^t\phi(t-\tau)\,\mathbf{B}\mathbf{u}(\tau)\,d\tau$ $\hfill (1.25)$

or, $\quad \mathbf{x}(t) = e^{\mathbf{A}(t-t_0)}\,\mathbf{x}(t_0) + \int_{t_0}^t e^{\mathbf{A}(t-\tau)}\,\mathbf{B}\mathbf{u}(\tau)\,d\tau.$ $\hfill (1.26)$

Once the state transition equation (time response) is determined, the output vector can easily be expressed as

$$y(t) = \mathbf{C}\,\phi(t-t_0)\mathbf{x}(t_0) + \int_{t_0}^t \mathbf{C}\,\phi\,(t-\tau)\,\mathbf{B}\mathbf{u}(\tau)\,d\tau + \mathbf{D}\mathbf{u}(t). \hfill (1.27)$$

Example 1.1. *Compute STM, when*

$$\mathbf{A} = \begin{bmatrix} 0 & 1 \\ -2 & -3 \end{bmatrix}.$$

Solution. $\qquad \mathbf{A} = \begin{bmatrix} 0 & 1 \\ -2 & -3 \end{bmatrix}$

$\therefore \qquad\qquad \left[s\mathbf{I} - \mathbf{A}\right] = \begin{bmatrix} s & 0 \\ 0 & s \end{bmatrix} - \begin{bmatrix} 0 & 1 \\ -2 & -3 \end{bmatrix} = \begin{bmatrix} s & -1 \\ 2 & s+3 \end{bmatrix}$

Now, $\quad \text{Adj } \mathbf{A} = \begin{bmatrix} s+3 & -2 \\ 1 & s \end{bmatrix}^T = \begin{bmatrix} s+3 & 1 \\ -2 & s \end{bmatrix}$

$\qquad\qquad \left[s\mathbf{I} - \mathbf{A}\right]^{-1} = \dfrac{\text{Adj}\,\mathbf{A}}{|s\mathbf{I} - \mathbf{A}|}$

$$= \frac{\begin{bmatrix} s+3 & 1 \\ -2 & s \end{bmatrix}}{\begin{vmatrix} s & -1 \\ 2 & s+3 \end{vmatrix}} = \frac{1}{(s^2 + 3s + 2)} \begin{bmatrix} s+3 & 1 \\ -2 & s \end{bmatrix}$$

$$= \begin{bmatrix} \dfrac{(s+3)}{(s+1)(s+2)} & \dfrac{1}{(s+1)(s+2)} \\ \dfrac{-2}{(s+1)(s+2)} & \dfrac{s}{(s+1)(s+2)} \end{bmatrix} = \begin{bmatrix} \left(\dfrac{2}{s+1} - \dfrac{1}{s+2} \right) & \left(\dfrac{1}{s+1} - \dfrac{1}{s+2} \right) \\ \left(\dfrac{-2}{s+1} + \dfrac{2}{s+2} \right) & \left(\dfrac{-1}{s+1} + \dfrac{2}{s+2} \right) \end{bmatrix}$$

$\therefore$ STM is given by

$$\phi(t) = \pounds^{-1}[sI - A]^{-1}$$

or, $$\phi(t) = \begin{bmatrix} (2e^{-t} - e^{-2t}) & (e^{-t} - e^{-2t}) \\ (-2e^{-t} + 2e^{-2t}) & (-e^{-t} + 2e^{-2t}) \end{bmatrix} = e^{At}$$ **Ans.**

Example 1.2. *A linear time-invariant system is characterised by the homogeneous state equation*

$$\begin{bmatrix} \dot{x}_1 \\ \dot{x}_2 \end{bmatrix} = \begin{bmatrix} 1 & 0 \\ 1 & 1 \end{bmatrix} \begin{bmatrix} x_1 \\ x_2 \end{bmatrix}.$$

Find the time response of the system. Assume the initial state vector

$$x(0) = \begin{bmatrix} 1 \\ 0 \end{bmatrix}.$$

Solution. $$\begin{bmatrix} \dot{x}_1 \\ \dot{x}_2 \end{bmatrix} = \begin{bmatrix} 1 & 0 \\ 1 & 1 \end{bmatrix} \begin{bmatrix} x_1 \\ x_2 \end{bmatrix}$$ (1.28)

Comparing Equation (1.28) with (1.19) we have

$$\mathbf{A} = \begin{bmatrix} 1 & 0 \\ 1 & 1 \end{bmatrix}$$

$$\therefore \qquad [s\mathbf{I} - \mathbf{A}] = \begin{bmatrix} s & 0 \\ 0 & s \end{bmatrix} - \begin{bmatrix} 1 & 0 \\ 1 & 1 \end{bmatrix} = \begin{bmatrix} s-1 & 0 \\ -1 & s-1 \end{bmatrix}$$

Now,

$$\text{Adj } \mathbf{A} = \begin{bmatrix} s-1 & 1 \\ 0 & s-1 \end{bmatrix}^{\mathrm{T}} = \begin{bmatrix} s-1 & 0 \\ 1 & s-1 \end{bmatrix}$$

$$[s\mathbf{I} - \mathbf{A}]^{-1} = \frac{\text{Adj } \mathbf{A}}{|s\mathbf{I} - \mathbf{A}|}$$

$$\therefore$$

$$= \frac{\begin{bmatrix} s-1 & 0 \\ 1 & s-1 \end{bmatrix}}{\begin{vmatrix} s-1 & 0 \\ -1 & s-1 \end{vmatrix}} = \frac{1}{(s-1)^2}\begin{bmatrix} s-1 & 0 \\ 1 & s-1 \end{bmatrix} = \begin{bmatrix} \dfrac{1}{(s-1)} & 0 \\ \dfrac{1}{(s-1)^2} & \dfrac{1}{(s-1)} \end{bmatrix}$$

$$\therefore \quad \text{STM is given by}$$

$$\phi(t) = \pounds^{-1}[s\mathbf{I} - \mathbf{A}]^{-1}$$

$$= \begin{bmatrix} e^t & 0 \\ te^t & e^t \end{bmatrix} = e^{\mathbf{A}t}.$$

Now, the time response of a linear, time-invariant homogeneous system is given by

$$\mathbf{x}(t) = \phi(t)\mathbf{x}(0)$$

$$= \begin{bmatrix} e^t & 0 \\ te^t & e^t \end{bmatrix}\begin{bmatrix} 1 \\ 0 \end{bmatrix} = \begin{bmatrix} e^t \\ te^t \end{bmatrix} \qquad\qquad \textbf{Ans.}$$

Example 1.3. *Obtain the time response of the following system*

$$\begin{bmatrix} \dot{x}_1 \\ \dot{x}_2 \end{bmatrix} = \begin{bmatrix} 1 & 0 \\ 1 & 1 \end{bmatrix} \begin{bmatrix} x_1 \\ x_2 \end{bmatrix} + \begin{bmatrix} 1 \\ 1 \end{bmatrix} u$$

where u(t) is a unit step occurring at $t = 0$ and $x^T(0) = [1\ 0]$.

Solution. We have

$$\mathbf{A} = \begin{bmatrix} 1 & 0 \\ 1 & 1 \end{bmatrix}, \mathbf{B} = \begin{bmatrix} 1 \\ 1 \end{bmatrix}.$$

The STM can be calculated as in Example 1.2.

We have

$$\phi(t) = \begin{bmatrix} e^t & 0 \\ te^t & e^t \end{bmatrix}.$$

The time response of the linear, time-invariant nonhomogeneous system is given by

$$\mathbf{x}(t) = \phi(t)\mathbf{x}(0) + \int_0^t \phi(t-\tau)\mathbf{B}u(\tau)d\tau$$

$$= \phi(t)\left\{\mathbf{x}(0) + \int_0^t \phi(-\tau)\mathbf{B}u(\tau)d\tau\right\}. \tag{1.29}$$

Now, with $u(t) = 1; t > 0$

we may write: $\phi(-\tau)\mathbf{B}u(\tau) = \begin{bmatrix} e^{-\tau} & 0 \\ -\tau e^{-\tau} & e^{-\tau} \end{bmatrix}\begin{bmatrix} 1 \\ 1 \end{bmatrix} = \begin{bmatrix} e^{-\tau} \\ e^{-\tau}(1-\tau) \end{bmatrix}$

$$\therefore \quad \int_0^t \phi(-\tau)\mathbf{B}u(\tau)d\tau = \begin{bmatrix} -e^{-\tau} \\ \tau e^{-\tau} \end{bmatrix}_0^t = \begin{bmatrix} 1-e^{-t} \\ te^{-t} \end{bmatrix}.$$

Now, from Equation (1.29) we get

$$\mathbf{x}(t) = \begin{bmatrix} e^t & 0 \\ te^t & e^t \end{bmatrix}\left\{\begin{bmatrix} 1 \\ 0 \end{bmatrix} + \begin{bmatrix} 1-e^{-t} \\ te^{-t} \end{bmatrix}\right\}$$

$$= \begin{bmatrix} e^t \\ te^t \end{bmatrix} + \begin{bmatrix} e^t-1 \\ te^t \end{bmatrix} = \begin{bmatrix} 2e^t-1 \\ 2te^t \end{bmatrix} \qquad \textbf{Ans.}$$

Example 1.4. *Consider a control system with state model*

$$\begin{bmatrix} \dot{x}_1 \\ \dot{x}_2 \end{bmatrix} = \begin{bmatrix} 0 & 1 \\ -2 & -3 \end{bmatrix} \begin{bmatrix} x_1 \\ x_2 \end{bmatrix} + \begin{bmatrix} 0 \\ 2 \end{bmatrix} u$$

$x_1(0) = 0$, $x_2(0) = 1$, $u = unit\ step$.

Compute the STM and find the state response.

Solution. We have

$$\mathbf{A} = \begin{bmatrix} 0 & 1 \\ -2 & -3 \end{bmatrix}, \quad \mathbf{B} = \begin{bmatrix} 0 \\ 2 \end{bmatrix}$$

and

$$\mathbf{x}(0) = \begin{bmatrix} 0 \\ 1 \end{bmatrix}.$$

The STM can be calculated as in Example 1.1. We have

$$\phi(t) = \begin{bmatrix} (2e^{-t} - e^{-2t}) & (e^{-t} - e^{-2t}) \\ (-2e^{-t} + 2e^{-2t}) & (-e^{-t} + 2e^{-2t}) \end{bmatrix}.$$

The time response of the linear, time-invariant nonhomogeneous system is given by

$$\mathbf{x}(t) = \phi(t)\mathbf{x}(0) + \int_0^t \phi(t-\tau)\mathbf{B}u(\tau)d\tau$$

$$= \phi(t)\mathbf{x}(0) + \phi(t)\int_0^t \phi(t-\tau)\mathbf{B}u(\tau)d\tau. \tag{1.30}$$

Now, with $u(t) = 1$; $t > 0$ we may write:

$$\phi(-\tau)\mathbf{B}u(\tau) = \begin{bmatrix} (2e^{\tau} - e^{2\tau}) & (e^{\tau} - e^{2\tau}) \\ (-2e^{\tau} + 2e^{2\tau}) & (-e^{\tau} + 2e^{\tau}) \end{bmatrix} \begin{bmatrix} 0 \\ 2 \end{bmatrix}$$

$$= \begin{bmatrix} (2e^{\tau} - 2e^{2\tau}) \\ (-2e^{\tau} + 4e^{2\tau}) \end{bmatrix}$$

$$\therefore \quad \int_0^t \phi(-\tau)\mathbf{B}u(\tau)d\tau = \begin{bmatrix} (2e^{\tau} - e^{2\tau}) \\ (-2e^{\tau} + 2e^{2\tau}) \end{bmatrix}_0^t = \begin{bmatrix} (2e^t - e^{2t}) \\ (-2e^t + 2e^{2t}) \end{bmatrix} - \begin{bmatrix} 2-1 \\ -2+2 \end{bmatrix}$$

$$= \begin{bmatrix} (2e^t - e^{2t} - 1) \\ (-2e^t + 2e^{2t}) \end{bmatrix}.$$

Now, from Equation (1.30) we get

$$x(t) = \begin{bmatrix} (2e^{-t} - e^{-2t}) & (e^{-t} - e^{-2t}) \\ (-2e^{-t} + 2e^{-2t}) & (-e^{-t} + 2e^{-2t}) \end{bmatrix} \begin{bmatrix} 0 \\ 1 \end{bmatrix}$$

$$+ \begin{bmatrix} (2e^{-t} - e^{-2t}) & (e^{-t} - e^{-2t}) \\ (-2e^{-t} + 2e^{-2t}) & (-e^{-t} + 2e^{-2t}) \end{bmatrix} \begin{bmatrix} (2e^{t} - e^{2t} - 1) \\ (-2e^{t} + 2e^{2t}) \end{bmatrix}$$

$$= \begin{bmatrix} (e^{-t} - e^{-2t}) \\ (-e^{-t} + 2e^{-2t}) \end{bmatrix} + \begin{bmatrix} (1 - 2e^{-t} + e^{-2t}) \\ (2e^{-t} - 2e^{-2t}) \end{bmatrix}$$

or, $$x(t) = \begin{bmatrix} (1 - e^{-t}) \\ e^{-t} \end{bmatrix}$$ **Ans.**

1.10 SOLUTION OF STATE EQUATIONS FOR LINEAR TIME-VARYING SYSTEMS

1.10.1 State Transition Matrix

The state equation of a linear time-varying system is given by

$$\dot{\mathbf{x}}(t) = \mathbf{A}(t)\mathbf{x}(t) + \mathbf{B}(t)\mathbf{u}(t).$$

For a homogeneous (unforced) system

$$\mathbf{u}(t) = 0$$

we have $$\dot{\mathbf{x}}(t) = \mathbf{A}(t)\mathbf{x}(t) \qquad (1.31)$$

where $$\mathbf{x}(t) = \begin{bmatrix} x_1(t) \\ x_2(t) \\ \vdots \\ x_n(t) \end{bmatrix}_{n \times 1}$$; State Vector

and, $\mathbf{A}(t) = n \times n$ matrix whose elements are continuous functions of t in the interval t_0 to t.

The solution of the linear time-varying homogeneous state Equation (1.31) is given by

$$\mathbf{x}(t) = \phi(t, t_0)\mathbf{x}(t_0) \tag{1.32}$$

where $\phi(t, t_0)$ is the $n \times n$ nonsingular matrix satisfying the matrix differential equation

$$\left. \begin{aligned} \dot{\phi}(t, t_0) &= A(t)\phi(t, t_0) \\ \phi(t_0, t_0) &= I \end{aligned} \right\} \tag{1.33}$$

where ϕ (t, t_0) is called the **State Transition Matrix** for the linear time-varying system described by Equation (1.31).

For time-varying systems, the state transition matrix depends upon both t and t_0 and **not on** the difference $t - t_0$. It is important to note, however, that the state transition matrix for a time-varying system cannot, in general, be given as a matrix exponential. The state transition matrix $\phi(t, t_0)$ is given by a matrix exponential **if, and only if**, $A(t)$ and $\int_{t_0}^{t} A(\tau)d\tau$ **commute**,

i.e., if, and only if, $\mathbf{A}(t)$ and $\int_{0}^{t} A(\tau)d\tau$ commute; the STM is given by

$$\phi(t, t_0) = e^{\int_{t_0}^{t} A(\tau)d\tau}.$$

Note that, if $\mathbf{A}(t)$ is a constant matrix or diagonal matrix, $\mathbf{A}(t)$ and $\int_{t_0}^{t} A(\tau)d\tau$ commute. However, if $\mathbf{A}(t)$ and $\int_{t_0}^{t} A(\tau)d\tau$ do not commute, there is no simple way to compute STM.

In order to compute $\phi(t, t_0)$ numerically, we have the following series expansion for $\phi(t, t_0)$

$$\phi(t, t_0) = I + \int_{t_0}^{t} A(\tau)d\tau + \int_{t_0}^{t} A(\tau_1) \left\{ \int_{t_0}^{t} A(\tau_2)d\tau_2 \right\} d\tau_1 + \ldots\ldots \tag{1.34}$$

Example 1.5. *Compute STM for the time-varying system, described by*

$$\begin{bmatrix} \dot{x}_1 \\ \dot{x}_2 \end{bmatrix} = \begin{bmatrix} 0 & 1 \\ 0 & t \end{bmatrix} \begin{bmatrix} x_1 \\ x_2 \end{bmatrix}$$

$$t_0 = 0.$$

Solution. $t_0 = 0$ and

$$A(t) = \begin{bmatrix} 0 & 1 \\ 0 & t \end{bmatrix}$$

$\therefore$
$$\int_0^t \mathbf{A}(\tau)d\tau = \int_0^t \begin{bmatrix} 0 & 1 \\ 0 & \tau \end{bmatrix}d\tau = \begin{bmatrix} 0 & \tau \\ 0 & \tau^2/2 \end{bmatrix}_0^t = \begin{bmatrix} 0 & t \\ 0 & t^2/2 \end{bmatrix}.$$

Again,

$$\int_0^t \mathbf{A}(\tau_1)\left\{\int_0^{\tau_1} \mathbf{A}(\tau_2)d\tau_2\right\}d\tau_1 = \int_0^t \begin{bmatrix} 0 & 1 \\ 0 & \tau_1 \end{bmatrix}\left\{\int_0^{\tau_1} \begin{bmatrix} 0 & 1 \\ 0 & \tau_2 \end{bmatrix}d\tau_2\right\}d\tau_1$$

$$= \int_0^t \begin{bmatrix} 0 & 1 \\ 0 & \tau_1 \end{bmatrix}\begin{bmatrix} 0 & \tau \\ 0 & \tau_1^2/2 \end{bmatrix}d\tau_1$$

$$= \int_0^t \begin{bmatrix} 0 & \tau_1^2/2 \\ 0 & \tau_1^3/2 \end{bmatrix}d\tau_1 = \begin{bmatrix} 0 & \tau_1^3/6 \\ 0 & \tau_1^4/8 \end{bmatrix}_0^t = \begin{bmatrix} 0 & t^3/6 \\ 0 & t^4/8 \end{bmatrix}.$$

We have STM

$$\phi(t,0) = \begin{bmatrix} 1 & 0 \\ 0 & 1 \end{bmatrix} + \begin{bmatrix} 0 & t \\ 0 & t^2/2 \end{bmatrix} + \begin{bmatrix} 0 & t^3/6 \\ 0 & t^4/6 \end{bmatrix} + \ldots\ldots$$

$$= \begin{bmatrix} 1 & (t + t^3/6 + \ldots\ldots) \\ 0 & (1 + t^2/2 + t^4/6 + \ldots\ldots) \end{bmatrix}$$ **Ans.**

1.10.1.1 Properties of State Transition Matrix

1. $\phi(t_0, t_0) = I$
2. $\phi(t_2, t_1)\phi(t_1, t_0) = \phi(t_2, t_0)$

Proof. Consider Equation (1.32); we have following results

$$\mathbf{x}(t_1) = \phi(t_1, t_0)\,\mathbf{x}(t_0), \tag{1.35 A}$$

$$\mathbf{x}(t_2) = \phi(t_2, t_0)\,\mathbf{x}(t_0). \tag{1.35 B}$$

However, $x(t_2)$ is also given by

$$\mathbf{x}(t_2) = \phi(t_2, t_1)\,\mathbf{x}(t_1). \tag{1.36 A}$$

Put $x(t_1)$ from Equation (1.35 A) in (1.36 A) and we have

$$\mathbf{x}(t_2) = \phi(t_2, t_1)\, \phi(t_1, t_0)\, \mathbf{x}(t_0).$$
(1.36 B)

Comparing Equation (1.35 B) and (1.36 B), we have

$$\phi(t_2, t_1)\phi(t_1, t_0) = \phi(t_2, t_0).$$
(1.37)

3. $\phi(t_1, t_0) = \phi^{-1}(t_0, t_1)$

Proof. From Equation (1.37) we have

$$\phi(t_1, t_0) = \phi^{-1}(t_2, t_1)\, \phi(t_2, t_0).$$

Put $t_2 = t_0$ in the previous equation and we get

$$\phi(t_1, t_0) = \phi^{-1}(t_0, t_1)\, \phi(t_0, t_0)$$
$$= \phi^{-1}(t_0, t_1) \qquad\qquad [\quad \phi(t_0, t_0) = \mathbf{I}.]$$

1.10.2 State Transition Equation

Consider the state equation of a linear, time-varying nonhomogeneous system

$$\dot{\mathbf{x}}(t) = \mathbf{A}(t)\, \mathbf{x}(t) + \mathbf{B}(t)\, \mathbf{u}(t).$$
(1.38)

Let the solution of Equation (1.38) be

$$\mathbf{x}(t) = \phi(t, t_0)\, \boldsymbol{\varepsilon}(t)$$
(1.39)

where $\phi(t, t_0)$ is an $n \times n$ matrix satisfying the equation

$$\dot{\phi}(t, t_0) = \mathbf{A}(t)\phi(t, t_0); \quad \phi(t_0, t_0) = \mathbf{I}.$$

Now,
$$\dot{\mathbf{x}}(t) = \frac{d}{dt}[\phi(t, t_0)\boldsymbol{\varepsilon}(t)]$$
$$= \dot{\phi}(t, t_0)\boldsymbol{\varepsilon}(t) + \phi(t, t_0)\dot{\boldsymbol{\varepsilon}}(t).$$
(1.40)

Using Equation (1.40) we get,

$$\dot{\mathbf{x}}(t) = \mathbf{A}(t)\phi(t,t_0)\boldsymbol{\varepsilon}(t) + \phi(t,t_0)\dot{\boldsymbol{\varepsilon}}(t).$$

Using Equation (1.39) we get,

$$\dot{\mathbf{x}}(t) = \mathbf{A}(t)\mathbf{x}(t) + \phi(t,t_0)\dot{\boldsymbol{\varepsilon}}(t). \tag{1.41}$$

Comparing Equation (1.38) and (1.41) we have,

$$\phi(t,t_0)\dot{\boldsymbol{\varepsilon}}(t) = \mathbf{B}(t)\mathbf{u}(t)$$

or, $\qquad \dot{\boldsymbol{\varepsilon}}(t) = \phi^{-1}(t,t_0)\mathbf{B}(t)\mathbf{u}(t).$

Integrate both sides with respect to t between the limits t_0 and t, and we have

$$\boldsymbol{\varepsilon}(t) - \boldsymbol{\varepsilon}(t_0) = \int_{t_0}^{t} \phi^{-1}(\tau,t_0)\mathbf{B}(\tau)\mathbf{u}(\tau)d\tau$$

or, $\qquad \boldsymbol{\varepsilon}(t) = \boldsymbol{\varepsilon}(t_0) + \int_{t_0}^{t} \phi(t_0,\tau)\mathbf{B}(\tau)\mathbf{u}(\tau)d\tau \qquad [\because \ \phi^{-1}(\tau,t_0) = \phi(t_0,\tau)].$

Now, put $t = t_0$ in Equation (1.39) and we get

$$\mathbf{x}(t_0) = \phi(t_0,t_0)\boldsymbol{\varepsilon}(t_0)$$

or, $\qquad \mathbf{x}(t_0) = \boldsymbol{\varepsilon}(t_0)$

Thus, $\qquad \boldsymbol{\varepsilon}(t) = \mathbf{x}(t_0) + \int_{t_0}^{t} \phi(t_0,\tau)\mathbf{B}(\tau)\mathbf{u}(\tau)d\tau. \qquad [\ \phi(t_0,t_0) = \mathbf{I}].$

Multiply both sides by $\phi(t, t_0)$ and we have

$$\phi(t,t_0)\boldsymbol{\varepsilon}(t) = \phi(t,t_0)\mathbf{x}(t_0) + \phi(t,t_0)\int_{t_0}^{t} \phi(t_0,\tau)\mathbf{B}(\tau)\mathbf{u}(\tau)d\tau$$

or, $\qquad \mathbf{x}(t) = \phi(t,t_0)\mathbf{x}(t_0) + \int_{t_0}^{t} \phi(t,\tau)\mathbf{B}(\tau)\mathbf{u}(\tau)d\tau, \tag{1.42}$

which is the desired equation.

1.11   TRANSFER MATRIX FROM THE STATE MODEL

The concept of the transfer matrix is an extension of that of the transfer function. Here we shall first obtain transfer function of a linear single-input-single-output control system and then transfer the matrix of a linear multiple-input-multiple-output control system using respective state and output equations.

1.11.1 Transfer Function

Consider the state model of a linear single-input-single-output system

$$\left.\begin{array}{l} \dot{\mathbf{x}}(t) = \mathbf{A}\mathbf{x}(t) + \mathbf{B}u(t); \ \ \text{State Equation} \\ y(t) = \mathbf{C}\mathbf{x}(t) + du\,(t); \ \ \text{Output Equation} \end{array}\right\}. \qquad (1.43)$$

Taking the Laplace transform of Equations (1.43) we have

$$s\mathbf{X}(s) - \mathbf{x}(0) = \mathbf{A}\mathbf{X}(s) + \mathbf{B}U(s) \qquad (1.44\ A)$$
$$Y(s) = \mathbf{C}\mathbf{X}(s) + dU(s). \qquad (1.44\ B)$$

Consider Equation (1.44 A)

$$[s\mathbf{I} - \mathbf{A}]\mathbf{X}(s) = \mathbf{B}U(s) + \mathbf{x}(0).$$

Assume zero initial conditions, i.e., x(0) = 0, we get

$$[s\mathbf{I} - \mathbf{A}]\mathbf{X}(s) = \mathbf{B}U(s)$$

or,
$$\mathbf{X}(s) = [s\mathbf{I} - \mathbf{A}]^{-1}\mathbf{B}U(s). \qquad (1.45)$$

Put X(s) from Equation (1.45) into (1.44 B) and we get

$$Y(s) = \mathbf{C}[s\mathbf{I} - \mathbf{A}]^{-1}\mathbf{B}U(s) + dU(s)$$

or,
$$\frac{Y(s)}{U(s)} = \mathbf{C}[s\mathbf{I} - \mathbf{A}]^{-1}\mathbf{B} + d \qquad (1.46)$$

$$G(s) = \mathbf{C}[s\mathbf{I} - \mathbf{A}]^{-1}\mathbf{B} + d.$$

Equation (1.46) gives the expression of **transfer function** of a linear single-input-single-output system.

1.11.2 Transfer Matrix

Consider the state model of a liner multi-input-multi-output system

$$\left.\begin{array}{l}\dot{\mathbf{x}}(t) = \mathbf{Ax}(t) + \mathbf{Bu}(t); \quad \text{State Equation} \\ \mathbf{y}(t) = \mathbf{Cx}(t) + \mathbf{Du}\,(t); \quad \text{Output Equation}\end{array}\right\}. \tag{1.47}$$

Take the Laplace transform of Equations (1.47) and we have

$$s\mathbf{X}(s) - \mathbf{x}(0) = \mathbf{AX}(s) + \mathbf{BU}(s) \tag{1.48 A}$$

$$\mathbf{Y}(s) = \mathbf{CX}(s) + \mathbf{DU}(s). \tag{1.48 B}$$

Consider Equation (1.48 A) assuming zero initial conditions, i.e., $x(0) = 0$, we get

$$[s\mathbf{I} - \mathbf{A}]\mathbf{X}(s) = \mathbf{BU}(s)$$

or,

$$\mathbf{X}(s) = [s\mathbf{I} - \mathbf{A}]^{-1}\mathbf{BU}(s) \tag{1.49}$$

Put $\mathbf{X}(s)$ from Equation (1.49) into (1.48 B) and we get

$$Y(s) = \mathbf{C}[s\mathbf{I} - \mathbf{A}]^{-1}\mathbf{BU}(s) + \mathbf{DU}(s)$$

or,

$$\frac{Y(s)}{\mathbf{U}(s)} = \mathbf{C}[s\mathbf{I} - \mathbf{A}]^{-1}\mathbf{B} + \mathbf{D}$$

or,

$$\mathbf{G}(s) = \mathbf{C}[s\mathbf{I} - \mathbf{A}]^{-1}\mathbf{B} + \mathbf{D}. \tag{1.50}$$

Equation (1.50) gives the expression of the **transfer matrix** of a linear multi-input-multi-output system, where transfer matrix $\mathbf{G}(s)$ relates the output $\mathbf{Y}(s)$ to the input $\mathbf{U}(s)$ as

$$Y(s) = \mathbf{G}(s)\mathbf{U}(s)$$

or, in an expanded form

$$\begin{bmatrix} Y_1(s) \\ Y_2(s) \\ \vdots \\ Y_p(s) \end{bmatrix}_{p\times 1} = \begin{bmatrix} G_{11}(s) & G_{12}(s) & \cdots\cdots & G_{1m}(s) \\ G_{21}(s) & G_{22}(s) & \cdots\cdots & G_{2m}(s) \\ \vdots & \vdots & \cdots\cdots & \vdots \\ G_{p1}(s) & G_{p2}(s) & \cdots\cdots & G_{pm}(s) \end{bmatrix}_{p\times m} \begin{bmatrix} U_1(s) \\ U_2(s) \\ \vdots \\ U_m(s) \end{bmatrix}_{m\times 1}.$$

The $(i, j)^{\text{th}}$ element $G_{ij}(s)$; $i = 1, 2, \ldots, p$; $j = 1, 2, \ldots m$ of $\mathbf{G}(s)$ is the transfer function relating the ith ouput to the jth input.

It is clear that the transfer function expression given by Equation (1.46) is a **special case** of the transfer matrix expression given by Equation (1.50).

Example 1.6. *Determine the transfer function for the system given below*

$$\begin{bmatrix} \dot{x}_1 \\ \dot{x}_2 \end{bmatrix} = \begin{bmatrix} 1 & -2 \\ 4 & -5 \end{bmatrix} \begin{bmatrix} x_1 \\ x_2 \end{bmatrix} + \begin{bmatrix} 2 \\ 1 \end{bmatrix} u$$

and

$$y = \begin{bmatrix} 1 & 1 \end{bmatrix} \begin{bmatrix} x_1 \\ x_2 \end{bmatrix}.$$

Solution. $\quad \mathbf{A} = \begin{bmatrix} 1 & -2 \\ 4 & -5 \end{bmatrix}$; $\mathbf{B} = \begin{bmatrix} 2 \\ 1 \end{bmatrix}$; $\mathbf{C} = \begin{bmatrix} 1 & 1 \end{bmatrix}$; $d = 0$

$$[s\mathbf{I} - \mathbf{A}] = \begin{bmatrix} s & 0 \\ 0 & s \end{bmatrix} - \begin{bmatrix} 1 & -2 \\ 4 & -5 \end{bmatrix} = \begin{bmatrix} (s-1) & 2 \\ -4 & (s+5) \end{bmatrix}$$

Now, $\quad \text{Adj}[s\mathbf{I} - \mathbf{A}] = \begin{bmatrix} (s+5) & 4 \\ -2 & (s-1) \end{bmatrix}^{\text{T}} = \begin{bmatrix} (s+5) & -2 \\ 4 & (s-1) \end{bmatrix}$

$\therefore \qquad [s\mathbf{I} - \mathbf{A}]^{-1} = \dfrac{\text{Adj}[s\mathbf{I} - \mathbf{A}]}{|s\mathbf{I} - \mathbf{A}|}$

$$= \frac{1}{(s^2 + 4s + 3)} \begin{bmatrix} (s+5) & -2 \\ 4 & (s-1) \end{bmatrix} = \begin{bmatrix} \dfrac{(s+5)}{(s+1)(s+3)} & \dfrac{-2}{(s+1)(s+3)} \\ \dfrac{4}{(s+1)(s+3)} & \dfrac{(s-1)}{(s+1)(s+3)} \end{bmatrix}.$$

The transfer function is given by

$$G(s) = \mathbf{C}[s\mathbf{I} - \mathbf{A}]^{-1}\mathbf{B}$$

$$= \begin{bmatrix} 1 & 1 \end{bmatrix} \begin{vmatrix} \dfrac{(s+5)}{(s+1)(s+3)} & \dfrac{-2}{(s+1)(s+3)} \\ \dfrac{4}{(s+1)(s+3)} & \dfrac{(s-1)}{(s+1)(s+3)} \end{vmatrix} \begin{bmatrix} 2 \\ 1 \end{bmatrix}$$

$$= \begin{bmatrix} 1 & 1 \end{bmatrix} \begin{bmatrix} \dfrac{2(s+5)-2}{(s+1)(s+3)} \\[2ex] \dfrac{8+(s-1)}{(s+1)(s+3)} \end{bmatrix} = \begin{bmatrix} 1 & 1 \end{bmatrix} \begin{bmatrix} \dfrac{(2s+8)}{(s+1)(s+3)} \\[2ex] \dfrac{(s+7)}{(s+1)(s+3)} \end{bmatrix}$$

$$= \dfrac{(2s+8)}{(s+1)(s+3)} + \dfrac{(s+7)}{(s+1)(s+3)}$$

or, $$G(s) = \dfrac{3(s+5)}{(s+1)(s+3)}$$ **Ans.**

Example 1.7. *Determine the transfer matrix for the system given below*

$$\begin{bmatrix} \dot{x}_1 \\ \dot{x}_2 \end{bmatrix} = \begin{bmatrix} 0 & 3 \\ -2 & -5 \end{bmatrix} \begin{bmatrix} x_1 \\ x_2 \end{bmatrix} + \begin{bmatrix} 1 & 1 \\ 1 & 1 \end{bmatrix} \mathbf{u}(t)$$

and

$$\mathbf{y} = \begin{bmatrix} 2 & 1 \\ 1 & 0 \end{bmatrix} \begin{bmatrix} x_1 \\ x_2 \end{bmatrix}.$$

Solution.

$$\mathbf{A} = \begin{bmatrix} 0 & 3 \\ -2 & -5 \end{bmatrix}; \mathbf{B} = \begin{bmatrix} 1 & 1 \\ 1 & 1 \end{bmatrix}; \mathbf{C} = \begin{bmatrix} 2 & 1 \\ 1 & 0 \end{bmatrix}; \mathbf{D} = 0$$

$$[s\mathbf{I} - \mathbf{A}] = \begin{bmatrix} s & 0 \\ 0 & s \end{bmatrix} - \begin{bmatrix} 0 & 3 \\ -2 & -5 \end{bmatrix} = \begin{bmatrix} s & -3 \\ 2 & (s+5) \end{bmatrix}$$

Now, $$\text{Adj}[s\mathbf{I} - \mathbf{A}] = \begin{bmatrix} (s+5) & -2 \\ 3 & s \end{bmatrix}^T = \begin{bmatrix} (s+5) & 3 \\ -2 & s \end{bmatrix}$$

$$\therefore \quad [s\mathbf{I} - \mathbf{A}]^{-1} = \dfrac{\text{Adj}[s\mathbf{I} - \mathbf{A}]}{|s\mathbf{I} - \mathbf{A}|} = \dfrac{1}{(s^2 + 5s + 6)} \begin{bmatrix} (s+5) & 3 \\ -2 & s \end{bmatrix}$$

$$= \begin{bmatrix} \dfrac{(s+5)}{(s+2)(s+3)} & \dfrac{3}{(s+2)(s+3)} \\[3ex] \dfrac{-2}{(s+2)(s+3)} & \dfrac{s}{(s+2)(s+3)} \end{bmatrix}$$

∴ $\mathbf{G}(s) = \mathbf{C}[s\mathbf{I} - \mathbf{A}]^{-1}\mathbf{B}$

$$\mathbf{G}(s) = \begin{bmatrix} 2 & 1 \\ 1 & 0 \end{bmatrix} \begin{bmatrix} \dfrac{(s+5)}{(s+2)(s+3)} & \dfrac{3}{(s+2)(s+3)} \\ \dfrac{-2}{(s+2)(s+3)} & \dfrac{s}{(s+2)(s+3)} \end{bmatrix} \begin{bmatrix} 1 & 1 \\ 1 & 1 \end{bmatrix}$$

$$= \begin{bmatrix} 2 & 1 \\ 1 & 0 \end{bmatrix} \begin{bmatrix} \dfrac{(s+8)}{(s+2)(s+3)} & \dfrac{(s+8)}{(s+2)(s+3)} \\ \dfrac{(s-2)}{(s+2)(s+3)} & \dfrac{(s-2)}{(s+2)(s+3)} \end{bmatrix}$$

or, $\mathbf{G}(s) = \begin{bmatrix} \dfrac{(3s+14)}{(s+2)(s+3)} & \dfrac{(3s+14)}{(s+2)(s+3)} \\ \dfrac{(s+8)}{(s+2)(s+3)} & \dfrac{(s+8)}{(s+2)(s+3)} \end{bmatrix}$ **Ans.**

1.12 CHARACTERISTIC EQUATION, EIGENVALUES, AND EIGENVECTORS

The **eigenvalues** of an n x n matrix $\mathbf{A}$ are the roots of the **characteristic equation** given by

$$|\lambda\mathbf{I} - \mathbf{A}| = 0. \tag{1.51}$$

The eigenvalues are sometimes called the **characteristic roots**.

Any nonzero vector $\mathbf{m}_i$; $i = 1, 2, \ldots\ldots n$ that satisfies the matrix equation

$$\left(\lambda_i\mathbf{I} - \mathbf{A}\right)\mathbf{m}_i = 0 \tag{1.52}$$

is known as the **eigenvector** of $\mathbf{A}$ associated with eigenvalue $\lambda_i; i = 1, 2, \ldots\ldots n$.

Example 1.8. *Consider a matrix*

$$\mathbf{A} = \begin{bmatrix} 3 & 4 \\ 2 & 1 \end{bmatrix}.$$

Determine the eigenvalues and eigenvectors of matrix **A**.

Solution. The characteristic equation of matrix **A** is given by

$$\left| \lambda \mathbf{I} - \mathbf{A} \right| = 0.$$

Now, $(\lambda \mathbf{I} - \mathbf{A}) = \begin{bmatrix} \lambda & 0 \\ 0 & \lambda \end{bmatrix} - \begin{bmatrix} 3 & 4 \\ 2 & 1 \end{bmatrix} = \begin{bmatrix} (\lambda - 3) & -4 \\ -2 & (\lambda - 1) \end{bmatrix}$

$\therefore$ $\left| \lambda \mathbf{I} - \mathbf{A} \right| = \begin{vmatrix} (\lambda - 3) & -4 \\ -2 & (\lambda - 1) \end{vmatrix} = 0$

or, $(\lambda - 3)(\lambda - 1) - 8 = 0$

or, $\lambda^2 - 3\lambda - \lambda + 3 - 8 = 0$

or, $\lambda^2 - 4\lambda - 5 = 0$

or, $\lambda^2 - 5\lambda + \lambda - 5 = 0$

or, $\lambda(\lambda - 5) + (\lambda - 5) = 0$

or, $(\lambda - 5)(\lambda + 1) = 0.$

Therefore, the eigenvalues of matrix **A** are

$$\lambda_1 = 5$$
$$\lambda_2 = -1$$

Ans.

Now, the eigenvector $\mathbf{m}_1$ associated with eigenvalue $\lambda_1 = 5$ is obtained by solving equation

$$(\lambda_1 \mathbf{I} - \mathbf{A})\mathbf{m}_1 = 0$$

or, $\left\{ \begin{bmatrix} 5 & 0 \\ 0 & 5 \end{bmatrix} - \begin{bmatrix} 3 & 4 \\ 2 & 1 \end{bmatrix} \right\} \begin{bmatrix} m_{11} \\ m_{21} \end{bmatrix} = 0$

or, $\begin{bmatrix} 2 & -4 \\ -2 & 4 \end{bmatrix} \begin{bmatrix} m_{11} \\ m_{21} \end{bmatrix} = 0$

or, $2m_{11} - 4m_{21} = 0$

and $-2m_{11} + 4m_{21} = 0.$

Select $m_{11} = 2$

$\therefore$ $m_{21} = 1$

$\therefore$ The eigenvector associated with eigenvalue $\lambda_1 = 5$ is

$$\mathbf{m}_1 = \begin{bmatrix} m_{11} \\ m_{21} \end{bmatrix} = \begin{bmatrix} 2 \\ 1 \end{bmatrix}$$ **Ans.**

Again, the eigenvector $\mathbf{m}_2$ associated with eigenvalue $\lambda_2 = -1$ is obtained by

$$(\lambda_2 \mathbf{I} - \mathbf{A})\mathbf{m}_2 = 0$$

or,

$$\left\{ \begin{bmatrix} -1 & 0 \\ 0 & -1 \end{bmatrix} - \begin{bmatrix} 3 & 4 \\ 2 & 1 \end{bmatrix} \right\} \begin{bmatrix} m_{12} \\ m_{22} \end{bmatrix} = 0$$

or,

$$\begin{bmatrix} -4 & -4 \\ -2 & -2 \end{bmatrix} \begin{bmatrix} m_{12} \\ m_{22} \end{bmatrix} = 0$$

or,

$$-4m_{12} - 4m_{22} = 0$$
$$-2m_{12} - 2m_{22} = 0.$$

Select, $m_{12} = 1$

$\therefore$ $m_{22} = -1$

$\therefore$ The eigenvector associated with eigenvalue $\lambda_2 = -1$ is

$$\mathbf{m}_2 = \begin{bmatrix} m_{12} \\ m_{22} \end{bmatrix} = \begin{bmatrix} 1 \\ -1 \end{bmatrix}$$ **Ans.**

1.13   STATE-SPACE REPRESENTATION OF CONTROL SYSTEMS

1.13.1 State-Space Representation Using Physical-Variables

State-variable formulation for a simple electrical system that is an **RLC network** is shown in Figure 1.5. If the initial conditions $v_c(0)$, $i_1(0)$, $i_2(0)$ and the input signal $e(t)$ for $t \geq 0$ are known; then the behavior of the electrical network is completely specified for $t \geq 0$. Therefore, initial conditions $v_c(0)$, $i_1(0)$, $i_2(0)$ together with the input signal $e(t)$ for $t \geq 0$ constitute the minimal information needed. It then follows that a natural selection of the state-variables would be

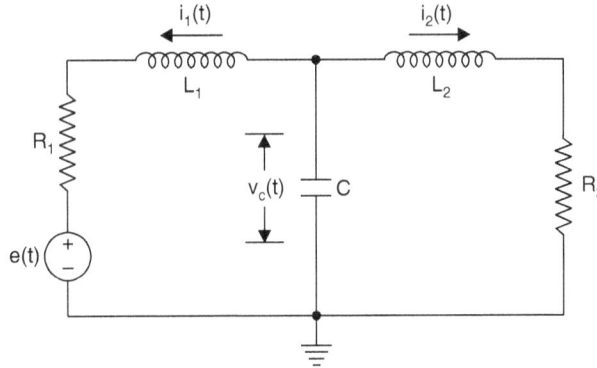

FIGURE 1.5 **An RLC network.**

$$x_1(t) = v_c(t), \qquad (1.53\ A)$$

$$x_2(t) = i_1(t), \qquad (1.53\ B)$$

$$x_3(t) = i_2(t). \qquad (1.53\ C)$$

However, the choice of the state variables for a given system is **not unique**.

Now, the differential equations, governing the behavior of the RLC network are

$$i_1(t) + i_2(t) + C\frac{dv_c(t)}{dt} = 0, \qquad (1.54\ A)$$

$$L_1\frac{di_1(t)}{dt} + R_1 i_1(t) + e(t) - v_c(t) = 0, \qquad (1.54\ B)$$

$$L_2\frac{di_2(t)}{dt} + R_2 i_2(t) - v_c(t) = 0. \qquad (1.54\ C)$$

We are interested in expressing the variables $\dfrac{dv_c(t)}{dt}, \dfrac{di_1(t)}{dt}$, and $\dfrac{di_2(t)}{dt}$ as linear combinations of the variables $v_c(t)$, $i_1(t)$, $i_2(t)$, and $e(t)$ as required in representation (1.10). For this purpose, Equations (1.54) may be rewritten as

$$\frac{dv_c(t)}{dt} = -\frac{1}{C} i_1(t) - \frac{1}{C} i_2(t)$$

$$\frac{di_1(t)}{dt} = \frac{1}{L_1} v_c(t) - \frac{R_1}{L_1} i_1(t) - \frac{1}{L_1} e(t)$$

$$\frac{di_2(t)}{dt} = \frac{1}{L_2} v_c(t) - \frac{R_2}{L_2} i_2(t)$$

or, in matrix form,

$$\begin{bmatrix} \dfrac{dv_c(t)}{dt} \\ \dfrac{di_1(t)}{dt} \\ \dfrac{di_2(t)}{dt} \end{bmatrix} = \begin{bmatrix} 0 & -\dfrac{1}{C} & -\dfrac{1}{C} \\ \dfrac{1}{L_1} & -\dfrac{R_1}{L_1} & 0 \\ \dfrac{1}{L_2} & 0 & -\dfrac{R_2}{L_2} \end{bmatrix} \begin{bmatrix} v_c(t) \\ i_1(t) \\ i_2(t) \end{bmatrix} + \begin{bmatrix} 0 \\ -\dfrac{1}{L_1} \\ 0 \end{bmatrix} e(t).$$

In terms of state-variables, defined in Equations (1.53) and the input $u(t) = e(t)$, we have the state-equations as

$$\begin{bmatrix} \dot{x}_1(t) \\ \dot{x}_2(t) \\ \dot{x}_3(t) \end{bmatrix} = \begin{bmatrix} 0 & -\dfrac{1}{C} & -\dfrac{1}{C} \\ \dfrac{1}{L_1} & -\dfrac{R_1}{L_1} & 0 \\ \dfrac{1}{L_2} & 0 & -\dfrac{R_2}{L_2} \end{bmatrix} \begin{bmatrix} x_1(t) \\ x_2(t) \\ x_3(t) \end{bmatrix} + \begin{bmatrix} 0 \\ -\dfrac{1}{L_1} \\ 0 \end{bmatrix} u(t). \qquad (1.55)$$

Let us assume the output-variables as

$$y_1(t) = v_{R_2}(t), \qquad (1.56\ A)$$

$$y_2(t) = i_{R_2}(t). \qquad (1.56\ B)$$

But $\qquad\qquad v_{R_2}(t) = R_2 i_2(t), \qquad (1.57\ A)$

$$i_{R_2}(t) = i_2(t). \qquad (1.57\ B)$$

Writing Equations (1.57) in matrix form, we get

$$\begin{bmatrix} v_{R_2}(t) \\ i_{R_2}(t) \end{bmatrix} = \begin{bmatrix} 0 & 0 & R_2 \\ 0 & 0 & 1 \end{bmatrix} \begin{bmatrix} v_c(t) \\ i_1(t) \\ i_2(t) \end{bmatrix}.$$

Thus, in terms of output-variables, defined in Equations (1.56) and state-variables, defined in Equations (1.53), we get the output equations as

$$\begin{bmatrix} y_1(t) \\ y_2(t) \end{bmatrix} = \begin{bmatrix} 0 & 0 & R_2 \\ 0 & 0 & 1 \end{bmatrix} \begin{bmatrix} x_1(t) \\ x_2(t) \\ x_3(t) \end{bmatrix}. \tag{1.58}$$

Equations (1.55) and (1.58) provide the state-model of the RLC network.

In the previous example, the selected state-variables are the physical quantities of the system, which can be measured. The choice of physical-variables of a system as state-variables, therefore, helps in implementation of the system-design.

Another advantage of selecting physical-variables for state-space representation is that the solution of state equations gives time-variation of variables, which have direct relevance to the physical system. However, with the choice of physical-variables, the solution of state equations may become a difficult task.

1.13.2 State-Space Representation Using Phase-Variables

The **phase-variables** are defined as those particular state-variables, which are obtained from one of the system-variables and its derivatives. Often, the variable used is the system output and the remaining state-variables are then derivatives of the output.

The phase-variable state model is easily determined if the system model is already known in the form of a differential equation/transfer function.

The **transfer function** of a general control system may be expressed as

$$G(s) = \frac{Y(s)}{U(s)} = \frac{b_0 s^m + b_1 s^{m-1} + \dots + b_{m-1} s + b_m}{s^n + a_1 s^{n-1} + \dots + a_{n-1} s + a_n}.$$

Most of the practical control schemes are realized with $m < n$. However, for the sake of generality, we consider the case that $m = n$. Thus,

$$G(s) = \frac{Y(s)}{U(s)} = \frac{b_0 s^n + b_1 s^{n-1} + \ldots + b_{n-1} s + b_n}{s^n + a_1 s^{n-1} + \ldots + a_{n-1} s + a_n}. \quad (1.59)$$

Using phase-variables, we can represent the system defined by Equation (1.59), in either of the phase-variable canonical forms discussed next.

1.13.2.1 Controllable Phase-Variable Canonical Form

Consider the transfer function system, defined by Equation (1.59). We may divide the transfer function $G(s)$ given by (1.59) into two parts as

$$G(s) = \frac{Y(s)}{U(s)} = \frac{Q(s)}{U(s)} \cdot \frac{Y(s)}{Q(s)} \quad (1.60)$$

where

$$\frac{Q(s)}{U(s)} = \frac{1}{s^n + a_1 s^{n-1} + \ldots + a_{n-1} s + a_n} \quad (1.61)$$

and,

$$\frac{Y(s)}{Q(s)} = b_0 s^n + b_1 s^{n-1} + \ldots + b_{n-1} s + b_n. \quad (1.62)$$

Equation (1.61) can be rewritten as

$$s^n Q(s) = -a_1 s^{n-1} Q(s) - \ldots - a_{n-1} s Q(s) - a_n Q(s) + U(s). \quad (1.63)$$

Define state-variables as follows

$$X_1(s) = Q(s)$$
$$X_2(s) = sQ(s)$$
$$\vdots \quad \vdots$$
$$X_{n-1}(s) = s^{n-2} Q(s)$$
$$X_n(s) = s^{n-1} Q(s).$$

Then, clearly

$$\left.\begin{aligned}
sX_1(s) &= sQ(s) = X_2(s)\\
sX_2(s) &= s^2 Q(s) = X_3(s)\\
&\vdots \quad \vdots \quad \vdots\\
sX_{n-1}(s) &= s^{n-1} Q(s) = X_n(s)\\
sX_n(s) &= s^n Q(s) = -a_1 s^{n-1} Q(s) - \ldots - a_{n-1} s Q(s) - a_n Q(s) + U(s)\\
&= -a_1 X_n(s) - \ldots - a_{n-1} X_2(s) - a_n X_1(s) + U(s)
\end{aligned}\right\}. \quad (1.64)$$

Taking the inverse Laplace transform of the set of Equations (1.64); assuming all initial conditions to be zero, we have

$$
\left.
\begin{aligned}
\dot{x}_1(t) &= x_2(t) \\
\dot{x}_2(t) &= x_3(t) \\
&\;\;\vdots \qquad \vdots \\
\dot{x}_{n-1}(t) &= x_n(t) \\
\dot{x}_n(t) &= -a_1 x_n(t) - \ldots\ldots - a_{n-1} x_2(t) - a_n x_1(t) + u(t)
\end{aligned}
\right\} \quad (1.65)
$$

The set of Equations (1.65) may be expressed in matrix form as

$$
\begin{bmatrix}
\dot{x}_1(t) \\
\dot{x}_2(t) \\
\vdots \\
\dot{x}_{n-1}(t) \\
\dot{x}_n(t)
\end{bmatrix}_{n\times 1}
=
\begin{bmatrix}
0 & 1 & 0 & \cdots & 0 \\
0 & 0 & 1 & \cdots & 0 \\
\vdots & \vdots & \vdots & \cdots & \vdots \\
0 & 0 & 0 & \cdots & 1 \\
-a_n & -a_{n-1} & -a_{n-2} & \cdots & -a_1
\end{bmatrix}_{n\times n}
\begin{bmatrix}
x_1(t) \\
x_2(t) \\
\vdots \\
x_{n-1}(t) \\
x_n(t)
\end{bmatrix}_{n\times 1}
+
\begin{bmatrix}
0 \\
0 \\
\vdots \\
0 \\
1
\end{bmatrix}_{n\times 1}
u(t)
\Bigg\} \quad (1.66)
$$

or,
$$
\dot{\mathbf{x}}(t) = \mathbf{A}\mathbf{x}(t) + \mathbf{B}u(t). \tag{1.67 A}
$$

Now, Equation (1.62) can be rewritten as

$$
Y(s) = b_0 s^n Q(s) + b_1 s^{n-1} Q(s) + \ldots\ldots + b_{n-1} s Q(s) + b_n Q(s).
$$

Putting the values from Equation (1.64) we get

$$
Y(s) = b_0 \{ -a_1 X_n(s) - \ldots\ldots - a_{n-1} X_2(s) - a_n X_1(s) + U(s) \} + b_1 X_n(s) + \cdots\cdots
$$
$$
+ b_{n-1} X_2(s) + b_n X_1(s)
$$

or,
$$
Y(s) = (b_n - a_n b_0) X_1(s) + (b_{n-1} - a_{n-1} b_0) X_2(s) + \ldots\ldots + (b_2 - a_2 b_0) X_{n-1}(s)
$$
$$
+ (b_1 - a_1 b_0) X_n(s) + b_0 U(s).
$$

Taking the inverse Laplace transform; assuming all initial conditions to be zero; we get

$$
y(t) = (b_n - a_n b_0) x_1(t) + (b_{n-1} - a_{n-1} b_0) x_2(t) + \ldots\ldots + (b_2 - a_2 b_0) x_{n-1}(t)
$$
$$
+ (b_1 - a_1 b_0) x_n(t) + b_0 u(t)
$$

or,   in matrix form $y(t) = \mathbf{C}x(t) + du(t)$ (1.67 B)

where, $\mathbf{C} = [(b_n - a_n b_0) \ (b_{n-1} - a_{n-1} b_0) \ .... \ (b_1 - a_1 b_0)]_{1 \times n}$

$$\mathbf{x}(t) = \begin{bmatrix} x_1(t) \\ x_2(t) \\ \vdots \\ x_{n-1}(t) \\ x_n(t) \end{bmatrix}_{n \times 1} ; \ \text{State Vector}$$

$d = b_0$; Transmission Constant

Equations (1.67) gives the phase-variable state model of the system in **controllable canonical form**. The block-diagram representation of the phase-variable state model in controllable canonical form is shown in Figure 1.6.

The matrix $\mathbf{A}$ in Equation (1.66) has a very special form having all 1s in the upper off-diagonal. Its last row comprises of the negative of the coefficients $a_n, a_{n-1}, \dots\dots a_1$ and all remaining elements are zero. This very special form of matrix $\mathbf{A}$ is known as the **Bush Form** or **Companion Form**. As the rank of matrix $\mathbf{A}$ is always 'n,' therefore, a system described by state Equations (1.67 A) is always controllable, hence, the name **controllable canonical form**. The concept of controllability shall be discussed later in this chapter.

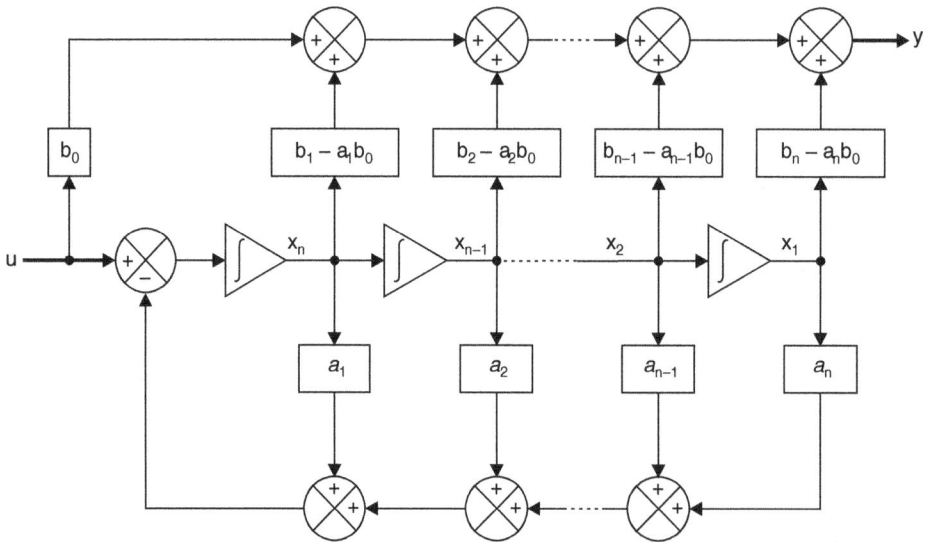

FIGURE 1.6 Block-diagram representation of the state model given by Equations (1.67).

1.13.2.2 Observable Phase-Variable Canonical Form

Consider the transfer function system, defined by Equation (1.59). It may be rewritten as

$$s^n Y(s) + a_1 s^{n-1} Y(s) + \ldots\ldots + a_{n-1} s Y(s) + a_n Y(s) = b_0 s^n U(s) + b_1 s^{n-1} U(s) + \cdots\cdots$$
$$+ b_{n-1} s U(s) + b_n U(s)$$

or,
$$Y(s) = b_0 U(s) + \frac{1}{s}[b_1 U(s) - a_1 Y(s)] + \ldots\ldots$$

$$+ \frac{1}{s^{n-1}}[b_{n-1} U(s) - a_{n-1} Y(s)] + \frac{1}{s^n}[b_n U(s) - a_n Y(s)] \quad (1.68)$$

Now, define the state-variables as follows

$$\left. \begin{aligned} X_n(s) &= \frac{1}{s}[b_1 U(s) - a_1 Y(s) + X_{n-1}(s)] \\[2mm] X_{n-1}(s) &= \frac{1}{s}[b_2 U(s) - a_2 Y(s) + X_{n-2}(s)] \\ \vdots \qquad \vdots \qquad \vdots \\ X_2(s) &= \frac{1}{s}[b_{n-1} U(s) - a_{n-1} Y(s) + X_1(s)] \\[2mm] X_1(s) &= \frac{1}{s}[b_n U(s) - a_n Y(s)] \end{aligned} \right\}. \qquad (1.69)$$

Then, Equation (1.68) may be written as

$$Y(s) = b_0 U(s) + X_n(s). \qquad (1.70)$$

Put $Y(s)$ from Equation (1.70) to Equation (1.69) and multiply both sides of the equations by s and we get

$$\left. \begin{aligned} sX_n(s) &= X_{n-1}(s) - a_1 X_n(s) + \left(b_1 - a_1 b_0\right) U(s) \\ sX_{n-1}(s) &= X_{n-2}(s) - a_2 X_n(s) + \left(b_2 - a_2 b_0\right) U(s) \\ \vdots \qquad \vdots \qquad \vdots \\ sX_2(s) &= X_1(s) - a_{n-1} X_n(s) + \left(b_{n-1} - a_{n-1} b_0\right) U(s) \\ sX_1(s) &= a_n X_n(s) + \left(b_n - a_n b_0\right) U(s) \end{aligned} \right\}. \qquad (1.71)$$

Taking the inverse Laplace transforms of the set of Equations (1.71); assuming all initial conditions to be zero; and writing them in the reverse order, we obtain

$$\left.\begin{array}{l} \dot{x}_1(t) = -a_n x_n(t) + (b_n - a_n b_0)u(t) \\ \dot{x}_2(t) = x_1(t) - a_{n-1}x_n(t) + (b_{n-1} - a_{n-1}b_0)u(t) \\ \vdots \qquad\qquad \vdots \qquad\qquad \vdots \\ \dot{x}_{n-1}(t) = x_{n-2}(t) - a_2 x_n(t) + (b_2 - a_2 b_0)u(t) \\ \dot{x}_n(t) = x_{n-1}(t) - a_1 x_n(t) + (b_1 - a_1 b_0)u(t) \end{array}\right\}. \qquad (1.72)$$

The set of Equations (1.72) may be expressed in matrix form as

$$\begin{bmatrix} \dot{x}_1(t) \\ \dot{x}_2(t) \\ \vdots \\ \dot{x}_{n-1}(t) \\ \dot{x}_n(t) \end{bmatrix}_{n\times1} = \begin{bmatrix} 0 & 0 & \cdots & 0 & 0 & -a_n \\ 1 & 0 & \cdots & 0 & 0 & -a_{n-1} \\ \vdots & \vdots & & \vdots & \vdots & \vdots \\ 0 & 0 & \cdots & 1 & 0 & -a_2 \\ 0 & 0 & \cdots & 0 & 1 & -a_1 \end{bmatrix}_{n\times n} \begin{bmatrix} x_1(t) \\ x_2(t) \\ \vdots \\ x_{n-1}(t) \\ x_n(t) \end{bmatrix}_{n\times1} + \begin{bmatrix} b_n - a_n b_0 \\ b_{n-1} - a_{n-1}b_0 \\ \vdots \\ b_2 - a_2 b_0 \\ b_1 - a_1 b_0 \end{bmatrix}_{n\times1} u(t) \qquad (1.73)$$

or, $\dot{\mathbf{x}}(t) = \mathbf{A}\mathbf{x}(t) + \mathbf{B}u(t).$ (1.74 A)

Also, taking the inverse Laplace transform of Equation (1.70), we get

$$y(t) = b_0 u(t) + x_n(t)$$

or, $$y(t) = \begin{bmatrix} 0 & 0 & \cdots\cdots & 0 & 1 \end{bmatrix}_{1\times n} \begin{bmatrix} x_1(t) \\ x_2(t) \\ \vdots \\ x_n(t) \end{bmatrix}_{n\times1} + b_0 u(t)$$

or, $y(t) = \mathbf{C}\mathbf{x}(t) + du(t).$ (1.74 B)

Equation (1.74) gives the phase-variable state model of the system in **observable canonical form**. The block-diagram representation of the phase-variable state model in observable canonical form is shown in Figure 1.7.

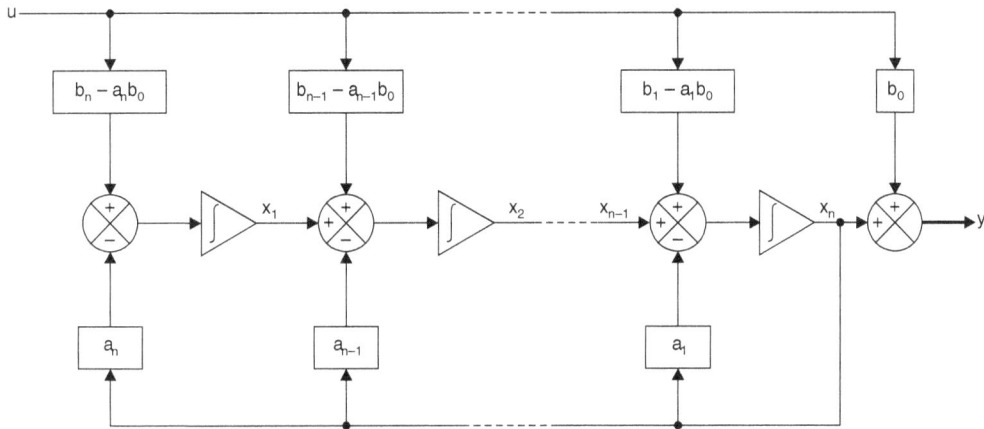

FIGURE 1.7 **Block-diagram representation of the state model given by Equations (1.74).**

The matrix **A** in Equation (1.73) has a very special form having all 1s in the lower off-diagonal. Its last column comprises the negative of coefficients a_n, a_{n-1}, ... a_1 and all remaining elements are zero. It can easily be verified that a system described by state Equation (1.74 A) is always observable, hence, the name **observable canonical form**. The concept of observability shall be discussed later in this chapter.

We observed that the phase-variable formulation can be obtained by inspection from the transfer function and **vice-versa**. A disadvantage of phase-variable formation is that the phase-variables, in general, are not physical (real) variables of the system and therefore are not available for **measurement** and **control purposes**; though phase-variables are simple to realize mathematically.

In spite of these disadvantages, phase-variables provide a powerful method of state-variable formulation. A **link** between the transfer function design approach and time-domain design approach is established through phase-variables.

Example 1.9. *Derive the controllable canonical form of the state-space representation for the transfer function system with poles only (i.e., it does not have zeros). Also represent it in block diagram.*

Solution. Consider Equation (1.59) and modify it so that the transfer function does not have zeros. The transfer function has the form

$$G(s) = \frac{Y(s)}{U(s)} = \frac{b_n}{s^n + a_1 s^{n-1} + \ldots\ldots + a_{n-1} s + a_n}. \tag{1.75}$$

Equation (1.75) can be rewritten as

$$s^n Y(s) = -a_1 s^{n-1} Y(s) - \ldots\ldots - a_{n-1} s Y(s) - a_n Y(s) + b_n U(s). \qquad (1.76)$$

Define state-variables as follows:

$$X_1(s) = Y(s)$$
$$X_2(s) = s Y(s)$$
$$\vdots \qquad \vdots$$
$$X_{n-1}(s) = s^{n-2} Y(s)$$
$$X_n(s) = s^{n-1} Y(s)$$

Then, clearly,

$$\left. \begin{aligned} s X_1(s) &= s Y(s) = X_2(s) \\ s X_2(s) &= s^2 Y(s) = X_3(s) \\ \vdots \quad &\vdots \quad \vdots \\ s X_{n-1}(s) &= s^{n-1} Y(s) = X_n(s) \\ s X_n(s) &= s^n Y(s) = -a_1 s^{n-1} Y(s) - \ldots\ldots - a_{n-1} s Y(s) - a_n Y(s) + b_n U(s) \\ &= -a_1 X_n(s) - \ldots\ldots - a_{n-1} X_2(s) - a_n X_1(s) + b_n U(s) \end{aligned} \right\} \qquad (1.77)$$

Taking the inverse Laplace transform of the set of Equations (1.77); assuming all initial conditions to be zero; we have

$$\left. \begin{aligned} \dot{x}(t) &= x_2(t) \\ \dot{x}_2(t) &= x_3(t) \\ \vdots \quad &\vdots \\ \dot{x}_{n-1}(t) &= x_n(t) \\ \dot{x}_n(t) &= -a_1 x_n(t) - \ldots\ldots - a_{n-1} x_2(t) - a_n x_1(t) + b_n u(t) \end{aligned} \right\} \qquad (1.78)$$

The set of Equations (1.78) may be expressed in matrix form as

$$\begin{bmatrix} \dot{x}_1(t) \\ \dot{x}_2(t) \\ \vdots \\ \dot{x}_{n-1}(t) \\ \dot{x}_n(t) \end{bmatrix}_{n\times 1} = \begin{bmatrix} 0 & 1 & 0 & \cdots & 0 \\ 0 & 0 & 1 & \cdots & 0 \\ \vdots & \vdots & \vdots & \cdots & \vdots \\ 0 & 0 & 0 & \cdots & 1 \\ -a_n & -a_{n-1} & -a_{n-2} & \cdots & -a_1 \end{bmatrix}_{n\times n} \begin{bmatrix} x_1(t) \\ x_2(t) \\ \vdots \\ x_{n-1}(t) \\ x_n(t) \end{bmatrix}_{n\times 1} + \begin{bmatrix} 0 \\ 0 \\ \vdots \\ 0 \\ b_n \end{bmatrix}_{n\times 1} u(t) \qquad (1.79)$$

or, $\qquad\qquad \dot{\mathbf{x}}(t) = \mathbf{A}\mathbf{x}(t) + \mathbf{B}u(t).$ $\qquad\qquad\qquad$ (1.80 A)

Also, $\qquad Y(s) = X_1(s).$

Taking the inverse Laplace transform

$$y(t) = x_1(t)$$

or, in matrix form,

$$y(t) = \mathbf{C}\mathbf{x}(t)$$

where, $\qquad \mathbf{C} = [1 \ \ 0 \ \ 0 \ldots\ldots 0]_1.$ $\qquad\qquad\qquad$ (1.80 B)

Equations (1.80) give the state-model of the system (in controllable canonical form), whose transfer function does not have zeros. The block-diagram representation of such a state model is shown in Figure 1.8.   **Ans.**

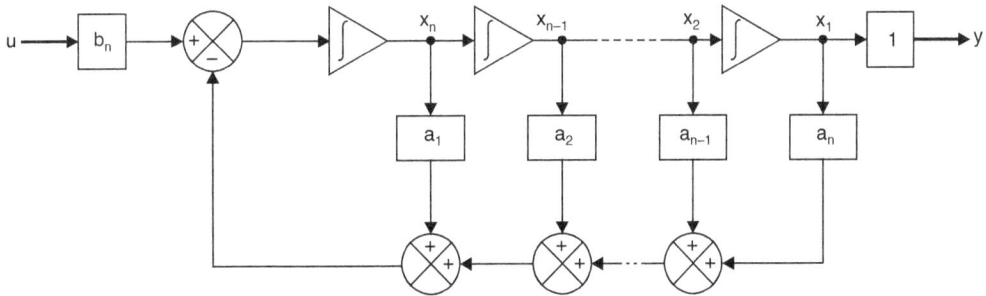

FIGURE 1.8 Block-diagram representation of the state model given by Equations (1.80).

Example 1.10. *A system is described by the following differential equation*

$$\frac{d^3 y}{dt^3} + 6\frac{d^2 y}{dt^2} + 11\frac{dy}{dt} + 6y = 2u. \qquad (1.81)$$

Obtain the state-space representation of the system. Also find the transfer function of the system.

Solution. Writing Equation (1.81) in s-domain, assuming all initial conditions to be zero, we have

$$s^3Y(s) + 6s^2Y(s) + 11sY(s) + 6Y(s) = 2U(s) \qquad (1.82)$$

or,

$$\frac{Y(s)}{U(s)} = \frac{2}{s^3 + 6s^2 + 11s + 6}$$

or,

$$G(s) = \frac{2}{s^3 + 6s^2 + 11s + 6}. \qquad (1.83)$$

Equation (1.83) gives the desired transfer function of the system. Again, from Equation (1.82), we have

$$s^3Y(s) = 2\,U(s) - 6s^2Y(s) - 11sY(s) - 6Y(s). \qquad (1.84)$$

Define state-variables as

$$X_1(s) = Y(s),$$
$$X_2(s) = sY(s),$$
$$X_3(s) = s^2Y(s).$$

Then clearly,

$$\left. \begin{array}{l} sX_1(s) = sY(s) = X_2(s) \\ sX_2(s) = s^2Y(s) = X_3(s) \\ sX_3(s) = s^3Y(s) = 2U(s) - 6s^2Y(s) - 11sY(s) - 6Y(s) \\ \qquad\qquad = 2U(s) - 6X_3(s) - 11X_2(s) - 6X_1(s) \end{array} \right\}. \qquad (1.85)$$

Taking the inverse Laplace transform of the set of Equations (1.85); assuming all initial conditions to be zero, we get

$$\dot{x}_1(t) = x_2(t)$$
$$\dot{x}_2(t) = x_3(t)$$
$$\dot{x}_3(t) = 2u(t) - 6x_3(t) - 11x_2(t) - 6x_1(t)$$

or,   in matrix form

$$\begin{bmatrix} \dot{x}_1(t) \\ \dot{x}_2(t) \\ \dot{x}_3(t) \end{bmatrix} = \begin{bmatrix} 0 & 1 & 0 \\ 0 & 0 & 1 \\ -6 & -11 & -6 \end{bmatrix} \begin{bmatrix} x_1(t) \\ x_2(t) \\ x_3(t) \end{bmatrix} + \begin{bmatrix} 0 \\ 0 \\ 2 \end{bmatrix} u(t).$$ (1.86 A)

Also, $Y(s) = X_1(s).$

Taking the inverse Laplace transform, we get

$$y(t) = x_1(t)$$

or, $$y(t) = \begin{bmatrix} 1 & 0 & 0 \end{bmatrix} \begin{bmatrix} x_1(t) \\ x_2(t) \\ x_3(t) \end{bmatrix}.$$ (1.86 B)

Equation (1.86) gives the desired state model of the system.   **Ans.**

Example 1.11. *Obtain the state model for the given transfer function.*

$$G(s) = \frac{Y(s)}{U(s)} = \frac{K(C_2 s + C_1)}{s^3 + a_3 s^2 + a_2 s + a_1}$$ (1.87)

Also show it in block-diagram.

Solution. Divide the transfer function $G(s)$ given by Equation (1.87) into two parts as

$$G(s) = \frac{Y(s)}{U(s)} = \frac{Q(s)}{U(s)} \cdot \frac{Y(s)}{Q(s)}$$

where $$\frac{Q(s)}{U(s)} = \frac{K}{s^3 + a_3 s^2 + a_2 s + a_1}$$ (1.88)

and, $$\frac{Y(s)}{Q(s)} = C_2 s + C_1.$$ (1.89)

Equation (1.88) can be rewritten as

$$s^3 Q(s) = -a_3 s^2 Q(s) - a_2 s Q(s) - a_1 Q(s) + KU(s).$$

Define state-variables as

$$X_1(s) = Q(s),$$
$$X_2(s) = sQ(s),$$
$$X_3(s) = s^2 Q(s).$$

Then, clearly,

$$\left. \begin{aligned} sX_1(s) &= sQ(s) = X_2(s) \\ sX_2(s) &= s^2 Q(s) = X_3(s) \\ sX_3(s) &= s^3 Q(s) = -a_3 s^2 Q(s) - a_2 sQ(s) - a_1 Q(s) + KU(s) \\ &= -a_3 X_3(s) - a_2 X_2(s) - a_1 X_1(s) + KU(s) \end{aligned} \right\}. \tag{1.90}$$

Taking the inverse Laplace transform of the set of Equations (1.90); assuming all initial conditions to be zero; we have

$$\dot{x}_1(t) = x_2 t$$
$$\dot{x}_2(t) = x_3(t)$$
$$\dot{x}_3(t) = -a_3 x_3(t) - a_2 x_2(t) - a_1 x_1(t) + Ku(t)$$

or,   in matrix form

$$\begin{bmatrix} \dot{x}_1(t) \\ \dot{x}_2(t) \\ \dot{x}_3(t) \end{bmatrix} = \begin{bmatrix} 0 & 1 & 0 \\ 0 & 0 & 1 \\ -a_1 & -a_2 & -a_3 \end{bmatrix} \begin{bmatrix} x_1(t) \\ x_2(t) \\ x_3(t) \end{bmatrix} + \begin{bmatrix} 0 \\ 0 \\ K \end{bmatrix} u(t). \tag{1.91 A}$$

Consider Equation (1.89), we have

$$Y(s) = C_2 sQ(s) + C_1 Q(s)$$
or,
$$Y(s) = C_2 X_2(s) + C_1 X_1(s).$$

Taking the inverse Laplace transform, we obtain

$$y(t) = C_2 x_2(t) + C_1 x_1(t)$$

or,   in matrix form

$$y(t) = [C_1 \quad C_2 \quad 0] \begin{bmatrix} x_1(t) \\ x_2(t) \\ x_3(t) \end{bmatrix}. \qquad (1.91\ B)$$

Equation (1.91) gives the desired state model of the system. The block-diagram representation of this state model is shown in Figure (1.9).

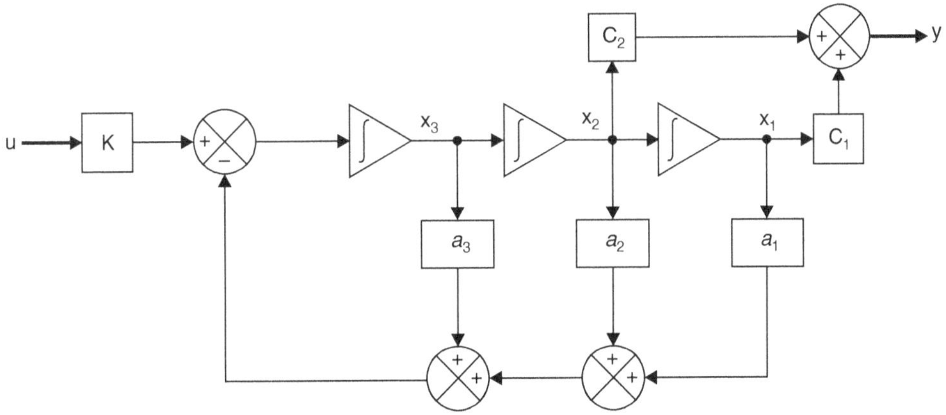

FIGURE 1.9   Block-diagram representation of the state model given in Equations (1.91).

1.13.3 State-Space Representation in Diagonal Canonical Form (Normal Form)

Consider the transfer function system, defined in Equation (1.59). Here, we consider the case where the denominator polynomial involves only **distinct roots.** For the **distinct-roots case,** Equation (1.59) can be rewritten as

$$G(s) = \frac{Y(s)}{U(s)} = \frac{b_0 s^n + b_1 s^{n-1} + \ldots\ldots + b_{n-1} s + b_n}{(s - \lambda_1)(s - \lambda_2)\ldots\ldots(s - \lambda_n)}$$

$$= b_0 + \frac{K_1}{s - \lambda_1} + \frac{K_2}{s - \lambda_2} + \ldots\ldots + \frac{K_n}{s - \lambda_n} \qquad (1.92)$$

where, K_i ($i = 1, 2, \ldots\ldots n$) are the residues of the poles at $s = \lambda_i$ ($i = 1, 2, \ldots\ldots, n$).

Equation (1.92) may be rewritten as

$$Y(s) = b_0 U(s) + \frac{K_1}{s - \lambda_1} U(s) + \frac{K_2}{s - \lambda_2} U(s) + \ldots\ldots + \frac{K_n}{s - \lambda_n} U(s). \qquad (1.93)$$

Define state-variables as follows

$$X_1(s) = \frac{1}{s - \lambda_1} U(s)$$

$$X_2(s) = \frac{1}{s - \lambda_2} U(s)$$

$$\vdots \qquad\qquad \vdots$$

$$X_n(s) = \frac{1}{s - \lambda_n} U(s)$$

Then, clearly,

$$sX_1(s) = \lambda_1 X_1(s) + U(s)$$
$$sX_2(s) = \lambda_2 X_2(s) + U(s)$$
$$\vdots \qquad \vdots \qquad \vdots$$
$$sX_n(s) = \lambda_n X_n(s) + U(s)$$

Taking the inverse Laplace transforms of the previous equations; assuming all initial conditions to be zero, we have

$$\dot{x}_1(t) = \lambda_1 x_1(t) + u(t)$$
$$\dot{x}_2(t) = \lambda_2 x_2(t) + u(t)$$
$$\vdots \qquad \vdots \qquad \vdots$$
$$\dot{x}_n(t) = \lambda_n x_n(t) + u(t)$$

or,　in matrix form

$$\begin{bmatrix} \dot{x}_1(t) \\ \dot{x}_2(t) \\ \vdots \\ \dot{x}_n(t) \end{bmatrix}_{n \times 1} = \begin{bmatrix} \lambda_1 & 0 & 0 & \ldots\ldots & 0 \\ 0 & \lambda_2 & 0 & \ldots\ldots & 0 \\ \vdots & \vdots & \vdots & \ldots\ldots & \vdots \\ 0 & 0 & 0 & \ldots\ldots & \lambda_n \end{bmatrix}_{n \times n} \begin{bmatrix} x_1(t) \\ x_2(t) \\ \vdots \\ x_n(t) \end{bmatrix}_{n \times 1} + \begin{bmatrix} 1 \\ 1 \\ \vdots \\ 1 \end{bmatrix}_{n \times 1} u(t).$$

$$(1.94\ A)$$

Now, in terms of state-variables, Equation (1.93) can be rewritten as

$$Y(s) = b_0 U(s) + K_1 X_1(s) + K_2 X_2(s) + \ldots + K_n X_n(s).$$

Taking the inverse Laplace transform, we have

$$y(t) = b_0 u(t) + K_1 x_1(t) + K_2 x_2(t) + \ldots + K_n x_n(t)$$

or, in matrix form

$$y(t) = [K_1 \quad K_2 \quad \ldots\ldots \quad K_n]_{1 \times n} \begin{bmatrix} x_1(t) \\ x_2(t) \\ \vdots \\ x_n(t) \end{bmatrix}_{n \times 1} + b_0 u(t). \qquad (1.94\ B)$$

Equations (1.94) give the state model of the system in **diagonal canonical form**. The block-diagram representation is shown in Figure 1.10.

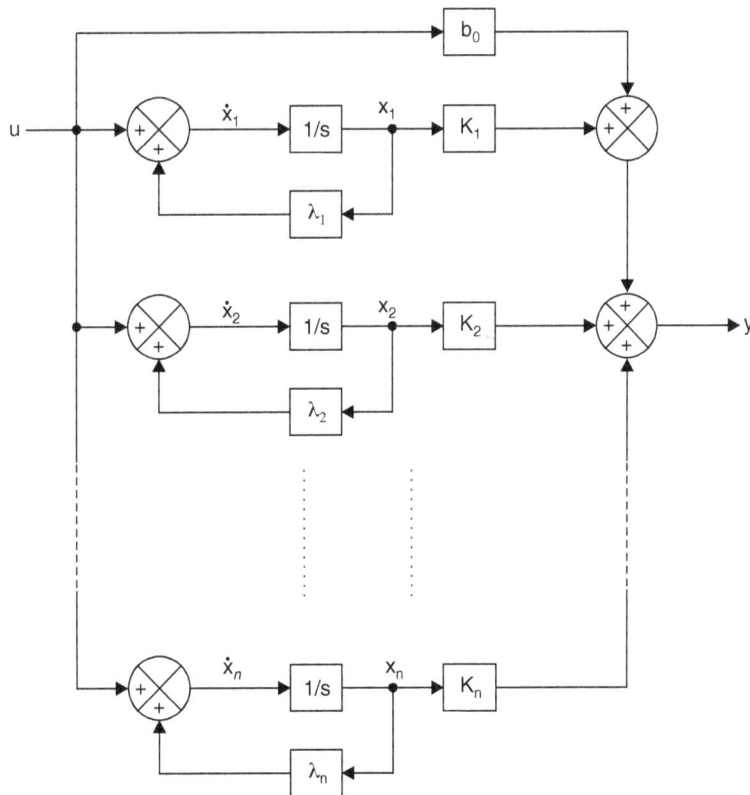

FIGURE 1.10 **Block-diagram representation of the system defined by Equations (1.94).**

An alternate state model in diagonal canonical form may be obtained simply by choosing the state-variables as follows

$$
\left.\begin{aligned}
X_1(s) &= \frac{K_1}{s - \lambda_1} U(s) \\
X_2(s) &= \frac{K_2}{s - \lambda_2} U(s) \\
&\vdots \qquad \vdots \\
X_n(s) &= \frac{K_n}{s - \lambda_n} U(s)
\end{aligned}\right\}. \tag{1.95}
$$

Then,   clearly,

$$
\begin{aligned}
sX_1(s) &= \lambda_1 X_1(s) + K_1 U(s) \\
sX_2(s) &= \lambda_2 X_2(s) + K_2 U(s) \\
&\vdots \qquad \vdots \qquad \vdots \\
sX_n(s) &= \lambda_n X_n(s) + K_n U(s)
\end{aligned}
$$

Taking the inverse Laplace transforms of the previous equations; assuming all initial conditions to be zero, we have

$$
\begin{aligned}
\dot{x}_1(t) &= \lambda_1 x_1(t) + K_1 u(t) \\
\dot{x}_2(t) &= \lambda_2 x_2(t) + K_2 u(t) \\
&\vdots \qquad \vdots \qquad \vdots \\
\dot{x}_n(t) &= \lambda_n x_n(t) + K_n u(t)
\end{aligned}
$$

or,   in matrix form

$$
\begin{bmatrix} \dot{x}_1(t) \\ \dot{x}_2(t) \\ \vdots \\ \dot{x}_n(t) \end{bmatrix}_{n \times 1} = \begin{bmatrix} \lambda_1 & 0 & 0 & \cdots & 0 \\ 0 & \lambda_2 & 0 & \cdots & 0 \\ \vdots & \vdots & \vdots & \cdots & \vdots \\ 0 & 0 & 0 & \cdots & \lambda_n \end{bmatrix}_{n \times n} \begin{bmatrix} x_1(t) \\ x_2(t) \\ \vdots \\ x_n(t) \end{bmatrix}_{n \times 1} + \begin{bmatrix} K_1 \\ K_2 \\ \vdots \\ K_n \end{bmatrix}_{n \times 1} u(t) \tag{1.96 A}
$$

Now, in terms of state-variables defined in Equation (1.95), Equation (1.93) can be rewritten as

$$
Y(s) = b_0 U(s) + X_1(s) + X_2(s) + \ldots + X_n(s).
$$

Taking the inverse Laplace transform, we get

$$y(t) = b_0 u(t) + x_1(t) + x_2(t) + \ldots\ldots + x_n(t)$$

or,   in matrix form

$$y(t) = \begin{bmatrix} 1 & 1 & \ldots\ldots & 1 \end{bmatrix}_{1 \times n} \begin{bmatrix} x_1(t) \\ x_2(t) \\ \vdots \\ x_n(t) \end{bmatrix}_{n \times 1} + b_0 u(t). \tag{1.96 B}$$

Equations (1.96) give the state model of the system in **diagonal canonical form**. The block diagram representation is shown in Figure 1.11.

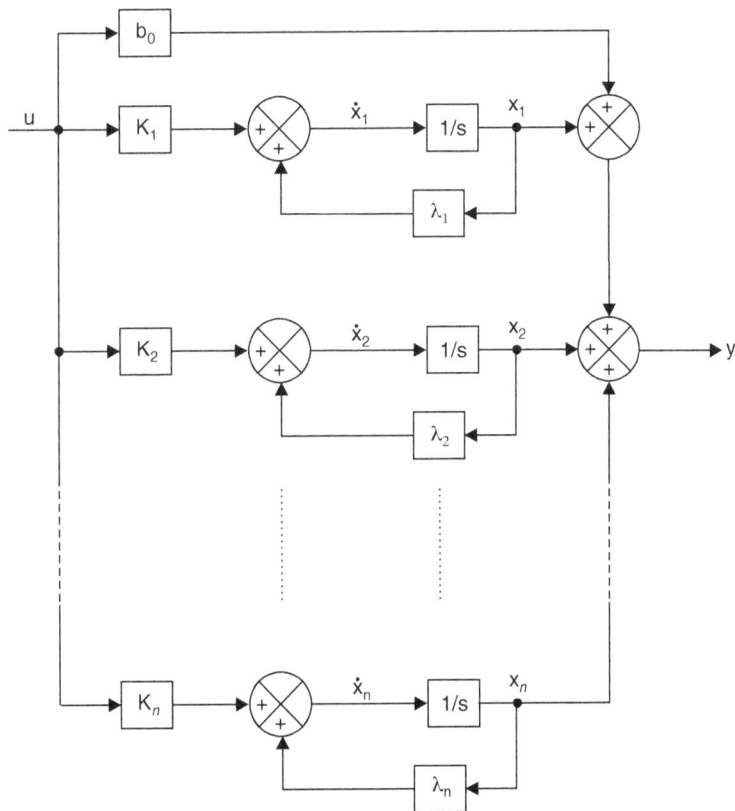

FIGURE 1.11 **Block-diagram representation of the system defined by Equations (1.96).**

In normal form or diagonal canonical form of the state model; the n first-order differential equations are completely independent of each other. This unique **decoupling feature** of normal form or diagonal canonical form greatly helps in the analysis of the system. Furthermore, the diagonal canonical form is sometimes also referred to as **canonical form** only.

1.13.4 State-Space Representation in Jordan Canonical Form

Consider the transfer function system, defined in Equation (1.59). Here, we consider the case where the denominator polynomial involves **multiple roots.** Let us assume that the denominator polynomial involves a **triple pole** at $s = \lambda_1$ and a **double pole** at $s = \lambda_4$. All other roots are **distinct** from each other. For this case, Equation (1.59) can be rewritten as

$$G(s) = \frac{Y(s)}{U(s)} = \frac{b_0 s^n + b_1 s^{n-1} + \ldots\ldots + b_{n-1} s + b_n}{(s - \lambda_1)^3 (s - \lambda_4)^2 (s - \lambda_6)(s - \lambda_7) \ldots\ldots (s - \lambda_n)}.$$

The partial fraction expansion becomes

$$\frac{Y(s)}{U(s)} = b_0 + \frac{K_1}{(s - \lambda_1)^3} + \frac{K_2}{(s - \lambda_1)^2} + \frac{K_3}{(s - \lambda_1)} + \frac{K_4}{(s - \lambda_4)^2} + \frac{K_5}{(s - \lambda_4)}$$

$$+ \frac{K_6}{(s - \lambda_6)} + \frac{K_7}{(s - \lambda_7)} + \ldots\ldots + \frac{K_n}{(s - \lambda_n)}$$

or,

$$Y(s) = b_0 U(s) + \frac{K_1}{(s - \lambda_1)^3} U(s) + \frac{K_2}{(s - \lambda_1)^2} U(s) + \frac{K_3}{(s - \lambda_1)} U(s) + \frac{K_4}{(s - \lambda_4)^2} U(s)$$

$$+ \frac{K_5}{(s - \lambda_4)} U(s) + \frac{K_6}{(s - \lambda_6)} U(s) + \frac{K_7}{(s - \lambda_7)} U(s) + \ldots\ldots + \frac{K_n}{(s - \lambda_n)} U(s).$$

$$(1.97)$$

Define, state-variables as

$$X_1(s) = \frac{1}{(s - \lambda_1)^3} U(s)$$

$$X_2(s) = \frac{1}{(s - \lambda_1)^2} U(s)$$

$$X_3(s) = \frac{1}{(s - \lambda_1)} U(s)$$

$$X_4(s) = \frac{1}{(s - \lambda_4)^2} U(s)$$

$$X_5(s) = \frac{1}{(s - \lambda_4)} U(s)$$

$$X_6(s) = \frac{1}{(s - \lambda_6)} U(s)$$

$$X_7(s) = \frac{1}{(s - \lambda_7)} U(s)$$

$$\vdots \qquad \vdots$$

$$X_n(s) = \frac{1}{(s - \lambda_n)} U(s)$$

Also, note that the following relationships exists among $X_1(s)$, $X_2(s)$, $X_3(s)$, and in between $X_4(s)$, $X_5(s)$

$$\frac{X_1(s)}{X_2(s)} = \frac{1}{(s - \lambda_1)},$$

$$\frac{X_2(s)}{X_3(s)} = \frac{1}{(s - \lambda_1)},$$

and,

$$\frac{X_4(s)}{X_5(s)} = \frac{1}{(s - \lambda_4)}.$$

Then, from the preceding definition of state-variables and the preceding relationships, we have

$$sX_1(s) = \lambda_1 X_1(s) + X_2(s)$$
$$sX_2(s) = \lambda_1 X_2(s) + X_3(s)$$
$$sX_3(s) = \lambda_1 X_3(s) + U(s)$$
$$sX_4(s) = \lambda_4 X_4(s) + X_5(s)$$
$$sX_5(s) = \lambda_4 X_5(s) + U(s)$$
$$sX_6(s) = \lambda_6 X_6(s) + U(s)$$

$$sX_7(s) = \lambda_7 X_7(s) + U(s)$$

$$\vdots \qquad \vdots \qquad \vdots$$

$$sX_n(s) = \lambda_n X_n(s) + U(s)$$

Taking the inverse Laplace transforms of the preceding equations; assuming all initial conditions to be zero; we have

$$\dot{x}_1(t) = \lambda_1 x_1(t) + x_2(t)$$
$$\dot{x}_2(t) = \lambda_1 x_2(t) + x_3(t)$$
$$\dot{x}_3(t) = \lambda_1 x_3(t) + u(t)$$
$$\dot{x}_4(t) = \lambda_4 x_4(t) + x_5(t)$$
$$\dot{x}_5(t) = \lambda_4 x_5(t) + u(t)$$
$$\dot{x}_6(t) = \lambda_6 x_6(t) + u(t)$$
$$\dot{x}_7(t) = \lambda_7 x_7(t) + u(t)$$
$$\vdots \qquad \vdots \qquad \vdots$$
$$\dot{x}_n(t) = \lambda_n x_n(t) + u(t)$$

or, in matrix form

$$
\begin{bmatrix} \dot{x}_1(t) \\ \dot{x}_2(t) \\ \dot{x}_3(t) \\ \dot{x}_4(t) \\ \dot{x}_5(t) \\ \dot{x}_6(t) \\ \dot{x}_7(t) \\ \vdots \\ \dot{x}_n(t) \end{bmatrix}_{n \times 1}
=
\begin{bmatrix}
\lambda_1 & 1 & 0 & 0 & 0 & 0 & 0 & \cdots & 0 \\
0 & \lambda_1 & 1 & 0 & 0 & 0 & 0 & \cdots & 0 \\
0 & 0 & \lambda_1 & 0 & 0 & 0 & 0 & \cdots & 0 \\
0 & 0 & 0 & \lambda_4 & 1 & 0 & 0 & \cdots & 0 \\
0 & 0 & 0 & 0 & \lambda_4 & 0 & 0 & \cdots & 0 \\
0 & 0 & 0 & 0 & 0 & \lambda_6 & 0 & \cdots & 0 \\
0 & 0 & 0 & 0 & 0 & 0 & \lambda_7 & \cdots & 0 \\
\vdots & \vdots & \vdots & \vdots & \vdots & \vdots & \vdots & & \vdots \\
0 & 0 & 0 & 0 & 0 & 0 & 0 & \cdots & \lambda_n
\end{bmatrix}_{n \times n}
\begin{bmatrix} x_1(t) \\ x_2(t) \\ x_3(t) \\ x_4(t) \\ x_5(t) \\ x_6(t) \\ x_7(t) \\ \vdots \\ x_n(t) \end{bmatrix}_{n \times 1}
+
\begin{bmatrix} 0 \\ 0 \\ 1 \\ 0 \\ 1 \\ 1 \\ 1 \\ \vdots \\ 1 \end{bmatrix}_{n \times 1}
u(t) .
$$

Jordan Block(s)

$$(1.98 \text{ A})$$

Now, in terms of state-variables, Equation (1.97) can be rewritten as

$$Y(s) = b_0 U(s) + K_1 X_1(s) + K_2 X_2(s) + K_3 X_3(s) + K_4 X_4(s) + \ldots\ldots + K_n X_n(s).$$

Taking the inverse Laplace transform, we get

$$y(t) = b_0 u(t) + K_1 x_1(t) + K_2 x_2(t) + K_3 x_3(t) + K_4 x_4(t) + \ldots\ldots + K_n x_n(t)$$

or,   in matrix form

$$y(t) = [K_1 \quad K_2 \quad K_3 \quad K_4 \quad \ldots\ldots \quad K_n]_{1 \times n} \begin{bmatrix} x_1(t) \\ x_2(t) \\ x_3(t) \\ x_4(t) \\ \vdots \\ x_n(t) \end{bmatrix}_{n \times 1} + b_0 u(t). \qquad (1.98\ B)$$

The state-space representation in the form given by Equations (1.98) is said to be in the **Jordan Canonical Form**. Figure 1.12 shows a block-diagram representation of the system given by Equations (1.98). The dotted sections in Equation (1.98 A) are called **Jordan Blocks**.

1.14   SIMILARITY TRANSFORMATION

The state equations of a system are **not unique** i.e., there exist more than one set of state-variables in terms of which the system behavior can be completely described. In fact, there are infinitely many state models for a given system and any two models are **uniquely related**. The transformation of one set of dynamic equations to another set of dynamic equations is termed as a '**similarity transformation.**'

Consider a multi-input-multi-output state model

$$\dot{\mathbf{x}}(t) = \mathbf{A}\mathbf{x}(t) + \mathbf{B}\mathbf{u}(t) \qquad (1.99\ A)$$
$$\mathbf{y}(t) = \mathbf{C}\mathbf{x}(t) + \mathbf{D}\mathbf{u}(t). \qquad (1.99\ B)$$

Let us define a new state-vector $\mathbf{z}(t)$ such that

$$\mathbf{x}(t) = \mathbf{T}\mathbf{z}(t)$$

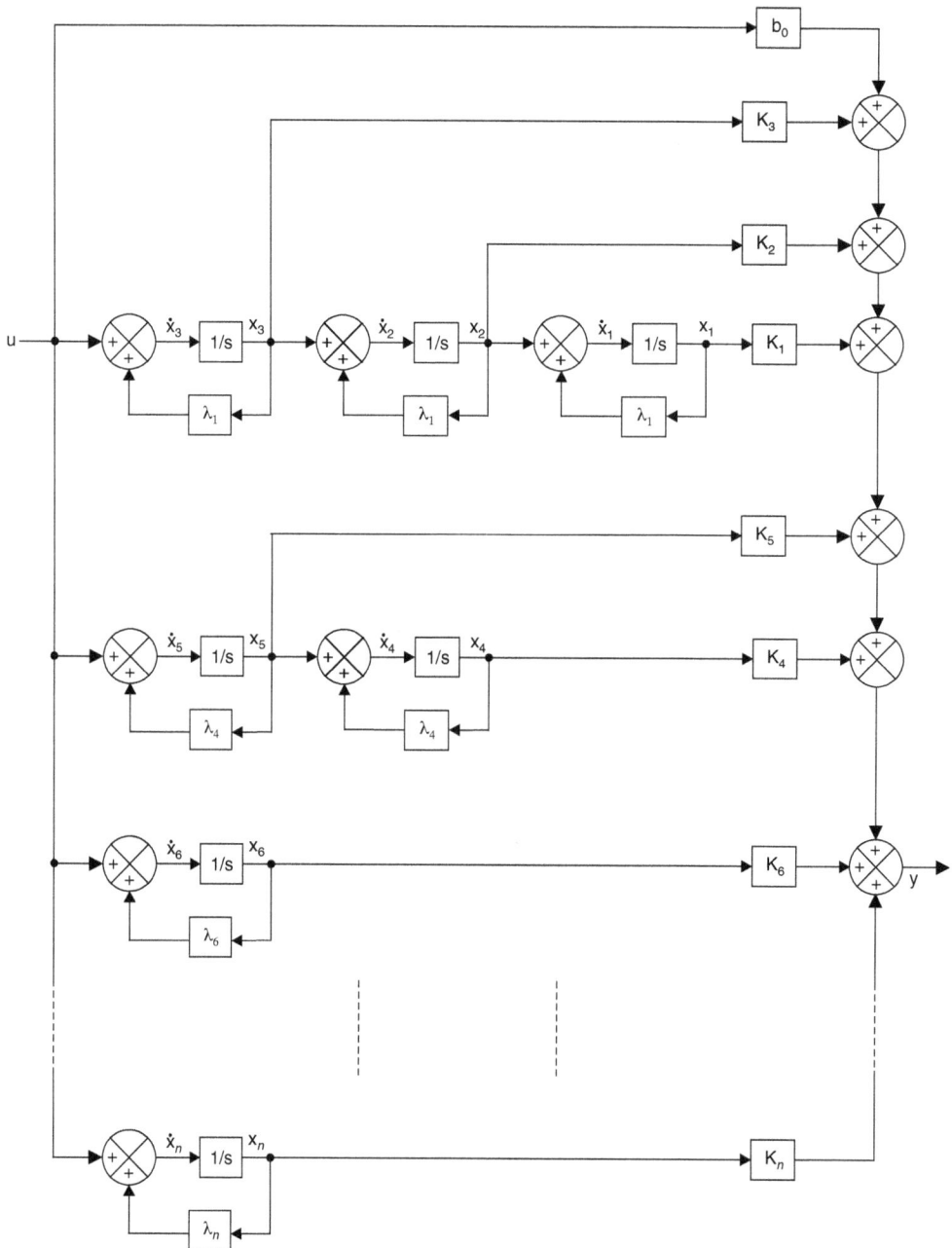

FIGURE 1.12   Block-diagram representation of the system defined by Equations (1.98).

where, $\mathbf{T}$ is an $n \times n$ nonsingular, constant matrix.

Since $\mathbf{T}$ is a constant matrix, it follows that

$$\dot{\mathbf{x}}(t) = \mathbf{T}\dot{\mathbf{z}}(t).$$

Substituting $\mathbf{x}(t)$ and $\dot{\mathbf{x}}(t)$ in Equation (1.99), we obtain

$$\mathbf{T}\dot{\mathbf{z}}(t) = \mathbf{A}\mathbf{T}\mathbf{z}(t) + \mathbf{B}\mathbf{u}(t) \qquad (1.100\ A)$$

$$\mathbf{y}(t) = \mathbf{C}\mathbf{T}\mathbf{z}(t) + \mathbf{D}\mathbf{u}(t). \qquad (1.100\ B)$$

Equations (1.100) may also be rewritten as

$$\dot{\mathbf{z}}(t) = \mathbf{T}^{-1}\mathbf{A}\mathbf{T}\mathbf{z}(t) + \mathbf{T}^{-1}\mathbf{B}\mathbf{u}(t) \qquad (1.101\ A)$$

$$\mathbf{y}(t) = \mathbf{C}\mathbf{T}\mathbf{z}(t) + \mathbf{D}\mathbf{u}(t). \qquad (1.101\ B)$$

Thus, the state model given by Equations (1.99) modifies to a new state model given by

$$\mathbf{z}(t) = \mathbf{A}\mathbf{z}(t) + \mathbf{B}\mathbf{u}(\text{t}) \qquad (1.102\ A)$$

$$\mathbf{y}(t) = \mathbf{C}\mathbf{z}(t) + \mathbf{D}\mathbf{u}(t) \qquad (1.102\ B)$$

where $\qquad\qquad \mathbf{A} = \mathbf{T}^{-1}\mathbf{A}\mathbf{T}$

$$\mathbf{B} = \mathbf{T}^{-1}\mathbf{B}$$

and, $\qquad\qquad \mathbf{C} = \mathbf{C}\mathbf{T}.$

Equations (1.102) give another state model for a given system. Since $\mathbf{T}$ is assumed to be a nonunique, nonsingular, constant matrix; the state model is also **nonunique**.

It is important to note that the transformation matrix $\mathbf{T}$ must be nonsingular, i.e.,

$$|\mathbf{T}| \neq 0$$

If this was not the case, it is obvious that the inverse transformation would not exist. The previous discussion, therefore, demonstrates that for a given system, infinitely many state models are possible and any two models are **uniquely related**.

1.14.1 Diagonalization

The physical-variable state model, in general, is not convenient for the investigation of system properties and evaluation of time-response. The diagonal canonical form, or normal form of state model, wherein the matrix $\mathbf{A}$ turns out to be a diagonal matrix, is most suitable for this purpose. Thus, it is useful to study the techniques by means of which a general state model can be transformed into a diagonal canonical form. These techniques are often referred to as **diagonalization techniques**.

Consider a multi-input-multi-output state model

$$\dot{\mathbf{x}}(t) = \mathbf{A}\mathbf{x}(t) + \mathbf{B}\mathbf{u}(t) \qquad (1.103\ A)$$

$$\mathbf{y}(t) = \mathbf{C}\mathbf{x}(t) + \mathbf{D}\mathbf{u}(t). \qquad (1.103\ B)$$

In the state model given by Equations (1.103), the matrix $\mathbf{A}$ is nondiagonal.

Let us define a new state-vector $\mathbf{z}(t)$ such that

$$\mathbf{x}(t) = \mathbf{M}\mathbf{z}(t)$$

where, $\mathbf{M}$ is an $(n \times n)$ nonsingular, constant matrix.

Under this similarity transformation, the original state model given by Equations (1.103) modifies to

$$\dot{\mathbf{z}}(t) = \mathbf{M}^{-1}\mathbf{A}\mathbf{M}\mathbf{z}(t) + \mathbf{M}^{-1}\mathbf{B}\mathbf{u}(t) \qquad (1.104\ A)$$

$$\mathbf{y}(t) = \mathbf{C}\mathbf{M}\mathbf{z}(t) + \mathbf{D}\mathbf{u}(t). \qquad (1.104\ B)$$

If we select the matrix $\mathbf{M}$ such that $\mathbf{M}^{-1}\mathbf{A}\mathbf{M}$ is a diagonalized matrix of matrix $\mathbf{A}$; then the model given by Equations (1.104) is a state model in diagonal canonical form. Under this condition, the matrix $\mathbf{M}$ is called the **diagonalizing matrix** or the **modal matrix**, and is constructed by placing the eigenvectors of matrix $\mathbf{A}$ together.

If $\mathbf{m}_1, \mathbf{m}_2, \mathbf{m}_3 \ldots, \mathbf{m}_n$ are the eigenvectors of matrix $\mathbf{A}$, corresponding to the eigenvalues $\lambda_1, \lambda_2, \lambda_3, \ldots, \lambda_n$ respectively, then the modal matrix is given by

$$\mathbf{M} = [\mathbf{m}_1 \vdots \mathbf{m}_2 \vdots \mathbf{m}_3 \vdots \cdots \mathbf{m}_n]_{n \times n}.$$

Thus, the general state model given by Equations (1.103) modifies to a new state model (in diagonal canonical form), given by

$$\dot{\mathbf{z}}(t) = \Lambda\mathbf{z}(t) + \tilde{\mathbf{B}}\mathbf{u}(t) \qquad (1.105\ A)$$

$$\mathbf{y}(t) = \tilde{\mathbf{C}}\mathbf{z}(t) + \mathbf{D}\mathbf{u}(t) \qquad (1.105\ B)$$

where $\qquad \Lambda = M^{-1}AM$; A Diagonal Matrix

$$\tilde{B} = M^{-1}B$$

and, $\qquad \tilde{C} = CM$

In fact, the matrix Λ could be obtained directly without the need to compute $M^{-1}AM$, since the diagonal elements of matrix Λ are given by distinct eigenvalues $\lambda_1, \lambda_2, \lambda_3, ..., \lambda_n$ of matrix A i.e.,

$$\Lambda = \begin{bmatrix} \lambda_1 & 0 & 0 & ... & 0 \\ 0 & \lambda_2 & 0 & ... & 0 \\ 0 & 0 & \lambda_3 & ... & 0 \\ \vdots & \vdots & \vdots & & \vdots \\ 0 & 0 & 0 & ... & \lambda_n \end{bmatrix}_{n \times n} = M^{-1}AM.$$

Note that the A and Λ matrices have the same characteristic equation; therefore, the eigenvalues are **invariant** under the transformation.

If matrix A is given in **Bush's** or **Phase-Variable or Companion Form** i.e.,

$$A = \begin{bmatrix} 0 & 1 & 0 & ... & 0 \\ 0 & 0 & 1 & ... & 0 \\ \vdots & \vdots & \vdots & & \vdots \\ 0 & 0 & 0 & ... & 1 \\ -a_n & -a_{n-1} & -a_{n-2} & ... & -a_1 \end{bmatrix}_{n \times n}$$

then, the modal matrix with reference to Bush's or Phase-Variable or Companion Form of matrix A can be shown to be a special matrix, called the **Vander Monde Matrix** and is given by

$$V = \begin{bmatrix} 1 & 1 & ... & 1 \\ \lambda_1 & \lambda_2 & ... & \lambda_n \\ \lambda_1^2 & \lambda_2^2 & ... & \lambda_n^2 \\ \vdots & \vdots & ... & \vdots \\ \lambda_1^{n-1} & \lambda_2^{n-1} & ... & \lambda_n^{n-1} \end{bmatrix}_{n \times n}$$

where $\lambda_1, \lambda_2, \lambda_3, ..., \lambda_n$ are distinct **eigenvalues** of matrix A.

An advantage of a diagonal matrix is that the inverse of such a matrix can be obtained merely by inspection, e.g.,

$$\begin{bmatrix} \alpha & 0 & 0 \\ 0 & \beta & 0 \\ 0 & 0 & \gamma \end{bmatrix}^{-1} = \begin{bmatrix} 1/\alpha & 0 & 0 \\ 0 & 1/\beta & 0 \\ 0 & 0 & 1/\gamma \end{bmatrix}.$$

Now consider the case in which the matrix **A** involves multiple eigenvalues– it is impossible to diagonalize. However, there exists a similarity transformation $\mathbf{x}(t) = \mathbf{S}\mathbf{z}(t)$ such that the matrix $\mathbf{J} = \mathbf{S}^{-1}\mathbf{A}\mathbf{S}$ is **almost a diagonal matrix**. The matrix **J** is called to be in **Jordan Canonical Form**. (Refer to Section 1.13.4). Under this similarity-transformation, the state model in Jordan Canonical Form can be expressed as

$$\dot{\mathbf{z}}(t) = \mathbf{J}\mathbf{z}(t) + \tilde{\mathbf{B}}\mathbf{u}(t) \tag{1.106 A}$$

$$\mathbf{y}(t) = \tilde{\mathbf{C}}\mathbf{z}(t) + \mathbf{D}\mathbf{u}(t) \tag{1.106 B}$$

where, $\mathbf{J} = \mathbf{S}^{-1}\mathbf{A}\mathbf{S}$; A Jordan Matrix (Almost Diagonal)

$$\tilde{\mathbf{B}} = \mathbf{S}^{-1}\mathbf{B}$$

and, $\tilde{\mathbf{C}} = \mathbf{C}\mathbf{S}$.

Note that, for each λ_i of **multiplicity** q, the Jordan matrix $\mathbf{J} = \mathbf{S}^{-1}\mathbf{A}\mathbf{S}$ will have a q x q **Jordan Block** corresponding to eigenvalue λ_i.

1.14.2 Computation of the State-Transition Matrix by Use of a Diagonal

In the previous section, we have seen that under the similarity transformation $\mathbf{x}(t) = \mathbf{M}\mathbf{z}(t)$, the matrix $\mathbf{\Lambda} = \mathbf{M}^{-1}\mathbf{A}\mathbf{M}$ is a diagonalized matrix of system matrix A with eigenvalues on its main diagonal.

Thus,

$$\mathbf{\Lambda} = \mathbf{M}^{-1}\mathbf{A}\mathbf{M}$$

or, $$e^{\mathbf{\Lambda}t} = \mathbf{M}^{-1}e^{\mathbf{A}t}\mathbf{M}$$

or, $$e^{\mathbf{A}t} = \mathbf{M}e^{\mathbf{\Lambda}t}\mathbf{M}^{-1}$$

or, $$\boldsymbol{\phi}(t) = \mathbf{M}e^{\mathbf{\Lambda}t}\mathbf{M}^{-1} \tag{1.107}$$

where, $\phi(t) = e^{At}$; State Transition Matrix.

Note that if the diagonalized matrix of **A** is given by

$$\Lambda = \begin{bmatrix} \lambda_1 & 0 & 0 \\ 0 & \lambda_2 & 0 \\ 0 & 0 & \lambda_3 \end{bmatrix}$$

then, $$e^{\Lambda t} = \begin{bmatrix} e^{\lambda_1 t} & 0 & 0 \\ 0 & e^{\lambda_2 t} & 0 \\ 0 & 0 & e^{\lambda_3 t} \end{bmatrix}.$$

In the case of **Jordan matrix J,** given by

$$J = \begin{bmatrix} \lambda_1 & 1 & 0 & 0 & 0 & 0 & 0 \\ 0 & \lambda_1 & 1 & 0 & 0 & 0 & 0 \\ 0 & 0 & \lambda_1 & 0 & 0 & 0 & 0 \\ 0 & 0 & 0 & \lambda_4 & 1 & 0 & 0 \\ 0 & 0 & 0 & 0 & \lambda_4 & 0 & 0 \\ 0 & 0 & 0 & 0 & 0 & \lambda_6 & 0 \\ 0 & 0 & 0 & 0 & 0 & 0 & \lambda_7 \end{bmatrix}$$

$$e^{Jt} = \begin{bmatrix} e^{\lambda_1 t} & te^{\lambda_1 t} & \frac{1}{2}t^2 e^{\lambda_1 t} & 0 & 0 & 0 & 0 \\ 0 & e^{\lambda_1 t} & te^{\lambda_1 t} & 0 & 0 & 0 & 0 \\ 0 & 0 & e^{\lambda_1 t} & 0 & 0 & 0 & 0 \\ 0 & 0 & 0 & e^{\lambda_4 t} & te^{\lambda_4 t} & 0 & 0 \\ 0 & 0 & 0 & 0 & e^{\lambda_4 t} & 0 & 0 \\ 0 & 0 & 0 & 0 & 0 & e^{\lambda_6 t} & 0 \\ 0 & 0 & 0 & 0 & 0 & 0 & e^{\lambda_7 t} \end{bmatrix}.$$

Example 1.12. *Consider a matrix A given by*

$$A = \begin{bmatrix} 0 & 1 & 0 \\ 3 & 0 & 2 \\ -12 & -7 & -6 \end{bmatrix}.$$

Compute the diagonal matrix Λ *and STM of* **A**.

Solution. The characteristic equation of matrix **A** is given by

$$|\lambda \mathbf{I} - \mathbf{A}| = 0.$$

Now,

$$\lambda \mathbf{I} - \mathbf{A} = \begin{bmatrix} \lambda & -1 & 0 \\ -3 & \lambda & -2 \\ 12 & 7 & \lambda+6 \end{bmatrix}$$

$\therefore$

$$|\lambda \mathbf{I} - \mathbf{A}| = \begin{bmatrix} \lambda & -1 & 0 \\ -3 & \lambda & -2 \\ 12 & 7 & \lambda+6 \end{bmatrix} = 0$$

or,

$$(\lambda + 1)(\lambda + 2)(\lambda + 3) = 0.$$

Therefore, the eigenvalues of matrix **A** are

$$\lambda_1 = -1,$$
$$\lambda_2 = -2,$$
$$\lambda_3 = -3.$$

Now, the eigenvector $\mathbf{m}_1$ associated with eigenvalue $\lambda_1 = -1$ is obtained by solving the equation

$$(\lambda_1 \mathbf{I} - \mathbf{A})\mathbf{m}_1 = 0$$

or,

$$\begin{bmatrix} -1 & -1 & 0 \\ -3 & -1 & -2 \\ 12 & 7 & 5 \end{bmatrix} \begin{bmatrix} m_{11} \\ m_{21} \\ m_{31} \end{bmatrix} = 0$$

$$-m_{11} - m_{21} = 0$$
$$-3m_{11} - m_{21} - 2m_{31} = 0$$
$$12m_{11} + 7m_{21} + 5m_{31} = 0.$$

Select $m_{11} = 1$, we get

$$m_{21} = -1$$

and, $m_{31} = -1$

thus, $$\mathbf{m}_1 = \begin{bmatrix} m_{11} \\ m_{21} \\ m_{31} \end{bmatrix} = \begin{bmatrix} 1 \\ -1 \\ -1 \end{bmatrix}.$$

Again, the eigenvector $\mathbf{m}_2$ associated with eigenvalue $\lambda_2 = -2$ is obtained by

$$(\lambda_2 \mathbf{I} - \mathbf{A})\mathbf{m}_2 = 0$$

or, $$\begin{bmatrix} -2 & -1 & 0 \\ -3 & -2 & -2 \\ 12 & 7 & 4 \end{bmatrix} \begin{bmatrix} m_{12} \\ m_{22} \\ m_{32} \end{bmatrix} = 0$$

or, $$-2m_{12} - m_{22} = 0$$
$$-3m_{12} - 2m_{22} - 2m_{32} = 0$$
$$12m_{12} + 7m_{22} + 4m_{32} = 0.$$

Select, $m_{12} = 2$, and we get

and, $$m_{22} = -4$$
$$m_{32} = 1$$

thus, $$\mathbf{m}_2 = \begin{bmatrix} m_{12} \\ m_{22} \\ m_{32} \end{bmatrix} = \begin{bmatrix} 2 \\ -4 \\ 1 \end{bmatrix}.$$

Similarly, we can compute the eigenvector $\mathbf{m}_3$ associated with $\lambda_3 = -3$, as

$$\mathbf{m}_3 = \begin{bmatrix} m_{13} \\ m_{23} \\ m_{33} \end{bmatrix} = \begin{bmatrix} 1 \\ -3 \\ 3 \end{bmatrix}.$$

The modal matrix **M** is obtained by

$$\mathbf{M} = [\mathbf{m}_1 \vdots \mathbf{m}_2 \vdots \mathbf{m}_3]$$

or,

$$\mathbf{M} = \begin{bmatrix} 1 & 2 & 1 \\ -1 & -4 & -3 \\ -1 & 1 & 3 \end{bmatrix}.$$

Now,

$$\text{Adj } \mathbf{M} = \begin{bmatrix} +\begin{vmatrix} -4 & -3 \\ 1 & 3 \end{vmatrix} & -\begin{vmatrix} -1 & -3 \\ -1 & 3 \end{vmatrix} & +\begin{vmatrix} -1 & -4 \\ -1 & 1 \end{vmatrix} \\ -\begin{vmatrix} 2 & 1 \\ 1 & 3 \end{vmatrix} & +\begin{vmatrix} 1 & 1 \\ -1 & 3 \end{vmatrix} & -\begin{vmatrix} 1 & 2 \\ -1 & 1 \end{vmatrix} \\ +\begin{vmatrix} 2 & 1 \\ -4 & -3 \end{vmatrix} & -\begin{vmatrix} 1 & 1 \\ -1 & -3 \end{vmatrix} & +\begin{vmatrix} 1 & 2 \\ -1 & -4 \end{vmatrix} \end{bmatrix}^T$$

$$= \begin{bmatrix} -9 & 6 & -5 \\ -5 & 4 & -3 \\ -2 & 2 & -2 \end{bmatrix}^T = \begin{bmatrix} -9 & -5 & -2 \\ 6 & 4 & 2 \\ -5 & -3 & -2 \end{bmatrix}.$$

Again,

$$|\mathbf{M}| = \begin{vmatrix} 1 & 2 & 1 \\ -1 & -4 & -3 \\ -1 & 1 & 3 \end{vmatrix} = -2$$

$$\therefore \quad \mathbf{M}^{-1} = \frac{\text{Adj } \mathbf{M}}{|\mathbf{M}|} = -\frac{1}{2}\begin{bmatrix} -9 & -5 & -2 \\ 6 & 4 & 2 \\ -5 & -3 & -2 \end{bmatrix}$$

or,

$$\mathbf{M}^{-1} = \begin{bmatrix} 4.5 & 2.5 & 1 \\ -3 & -2 & -1 \\ 2.5 & 1.5 & 1 \end{bmatrix}$$

$$\mathbf{\Lambda} = \mathbf{M}^{-1}\mathbf{A}\mathbf{M}$$

$$= \begin{bmatrix} 4.5 & 2.5 & 1 \\ -3 & -2 & -1 \\ 2.5 & 1.5 & 1 \end{bmatrix}\begin{bmatrix} 0 & 1 & 0 \\ 3 & 0 & 2 \\ -12 & -7 & -6 \end{bmatrix}\begin{bmatrix} 1 & 2 & 1 \\ -1 & -4 & -3 \\ -1 & 1 & 3 \end{bmatrix}$$

$$= \begin{bmatrix} 4.5 & 2.5 & 1 \\ -3 & -2 & -1 \\ 2.5 & 1.5 & 1 \end{bmatrix} \begin{bmatrix} -1 & -4 & -3 \\ 1 & 8 & 9 \\ 1 & -2 & -9 \end{bmatrix} = \begin{bmatrix} -1 & 0 & 0 \\ 0 & -2 & 0 \\ 0 & 0 & -3 \end{bmatrix} \quad \textbf{Ans.}$$

$$\therefore \; e^{\Lambda t} = \begin{bmatrix} e^{-t} & 0 & 0 \\ 0 & e^{-2t} & 0 \\ 0 & 0 & e^{-3t} \end{bmatrix}$$

The STM is given by

$$\boldsymbol{\phi}(t) \,=\, e^{At} \,=\, \mathbf{M} e^{\Lambda t}\, \mathbf{M}^{-1}$$

or,
$$\boldsymbol{\phi}(t) = \begin{bmatrix} 1 & 2 & 1 \\ -1 & -4 & -3 \\ -1 & 1 & 3 \end{bmatrix} \begin{bmatrix} e^{-t} & 0 & 0 \\ 0 & e^{-2t} & 0 \\ 0 & 0 & e^{-3t} \end{bmatrix} \begin{bmatrix} 4.5 & 2.5 & 1 \\ -3 & -2 & -1 \\ 2.5 & 1.5 & 1 \end{bmatrix}$$

or,
$$\boldsymbol{\phi}(t) = \begin{bmatrix} 1 & 2 & 1 \\ -1 & -4 & -3 \\ -1 & 1 & 3 \end{bmatrix} \begin{bmatrix} 4.5e^{-t} & 2.5e^{-t} & e^{-t} \\ -3e^{-2t} & -2e^{-2t} & -e^{-2t} \\ 2.5e^{-3t} & 1.5e^{-3t} & e^{-3t} \end{bmatrix}$$

or,
$$\boldsymbol{\phi}(t) = \begin{bmatrix} (4.5e^{-t} - 6e^{-2t} + 2.5e^{-3t}) & (2.5e^{-t} - 4e^{-2t} + 1.5e^{-3t}) & (e^{-t} + 2e^{-2t} + e^{-3t}) \\ (-4.5e^{-t} + 12e^{-2t} - 7.5e^{-3t}) & (-2.5e^{-t} + 8e^{-2t} - 4.5e^{-3t}) & (-e^{-t} + 4e^{-2t} - 3e^{-3t}) \\ (-4.5e^{-t} - 3e^{-2t} + 7.5e^{-3t}) & (-2.5e^{-t} - 2e^{-2t} + 4.5e^{-3t}) & (-e^{-t} - e^{-2t} + 3e^{-3t}) \end{bmatrix}$$

Example 1.13. *Consider the state-space representation of a system*

$$\begin{bmatrix} \dot{x}_1 \\ \dot{x}_2 \\ \dot{x}_3 \end{bmatrix} = \begin{bmatrix} 0 & 1 & 0 \\ 0 & 0 & 1 \\ -6 & -11 & -6 \end{bmatrix} \begin{bmatrix} x_1 \\ x_2 \\ x_3 \end{bmatrix} + \begin{bmatrix} 0 \\ 0 \\ 2 \end{bmatrix} u \qquad (1.108\ A)$$

$$y = \begin{bmatrix} 1 & 0 & 0 \end{bmatrix} \begin{bmatrix} x_1 \\ x_2 \\ x_3 \end{bmatrix}. \qquad (1.108\ B)$$

Change it into DCF (Diagonal Canonical Form).

Solution. Compare Equations (1.108) with

$$\dot{\mathbf{x}}(t) = \mathbf{A}\mathbf{x}(t) + \mathbf{B}u(t)$$
$$y(t) = \mathbf{C}\mathbf{x}(t).$$

We have

$$\mathbf{A} = \begin{bmatrix} 0 & 1 & 0 \\ 0 & 0 & 1 \\ -6 & -11 & -6 \end{bmatrix}, \mathbf{B} = \begin{bmatrix} 0 \\ 0 \\ 2 \end{bmatrix}$$

and,

$$\mathbf{C} = [1 \ 0 \ 0].$$

The characteristic equation of matrix **A** is given by

$$| \lambda\mathbf{I} - \mathbf{A} | = 0.$$

Now,

$$\lambda\mathbf{I} - \mathbf{A} = \begin{bmatrix} \lambda & 0 & 0 \\ 0 & \lambda & 0 \\ 0 & 0 & \lambda \end{bmatrix} - \begin{bmatrix} 0 & 1 & 0 \\ 0 & 0 & 1 \\ -6 & -11 & -6 \end{bmatrix}$$

$$= \begin{bmatrix} \lambda & -1 & 0 \\ 0 & \lambda & -1 \\ 6 & 11 & \lambda + 6 \end{bmatrix}$$

$$\therefore \qquad | \lambda\mathbf{I} - \mathbf{A} | = 0$$

or, $\qquad \lambda^3 + 6\lambda^2 + 11\lambda + 6 = 0$

or, $\qquad (\lambda + 1)(\lambda + 2)(\lambda + 3) = 0.$

Therefore, the eigenvalues of matrix **A** are

$$\lambda_1 = -1,$$
$$\lambda_2 = -2,$$
$$\lambda_3 = -3.$$

Matrix **A** is in Bush's form. Thus, in this case, the diagonalizing or Modal matrix will be the Vander Monde matrix, given as

$$\mathbf{V} = \begin{bmatrix} 1 & 1 & 1 \\ \lambda_1 & \lambda_2 & \lambda_3 \\ \lambda_1^2 & \lambda_2^2 & \lambda_3^2 \end{bmatrix} = \begin{bmatrix} 1 & 1 & 1 \\ -1 & -2 & -3 \\ 1 & 4 & 9 \end{bmatrix}$$

$$\therefore \qquad \mathbf{V}^{-1} = \begin{bmatrix} 3 & 2.5 & 0.5 \\ -3 & -4 & -1 \\ 1 & 1.5 & 0.5 \end{bmatrix}.$$

(It is left as an exercise for the readers to calculate $\mathbf{V}^{-1}$ themselves.)

Using similarity transformation $\mathbf{x}(t) = \mathbf{V}\mathbf{z}(t)$, the original state model modifies to diagonal canonical form

$$\dot{\mathbf{z}}(t) = \mathbf{\Lambda}\mathbf{z}(t) + \tilde{\mathbf{B}}u(t)$$
$$y(t) = \tilde{\mathbf{C}}\mathbf{z}(t)$$

where,
$$\mathbf{\Lambda} = \mathbf{V}^{-1}\mathbf{A}\mathbf{V}$$
$$\tilde{\mathbf{B}} = \mathbf{V}^{-1}\mathbf{B}$$
$$\tilde{\mathbf{C}} = \mathbf{C}\mathbf{V}.$$

Now,
$$\mathbf{\Lambda} = \mathbf{V}^{-1}\mathbf{A}\mathbf{V}$$

$$= \begin{bmatrix} 3 & 2.5 & 0.5 \\ -3 & -4 & -1 \\ 1 & 1.5 & 0.5 \end{bmatrix} \begin{bmatrix} 0 & 1 & 0 \\ 0 & 0 & 1 \\ -6 & -11 & -6 \end{bmatrix} \begin{bmatrix} 1 & 1 & 1 \\ -1 & -2 & -3 \\ 1 & 4 & 9 \end{bmatrix}$$

$$= \begin{bmatrix} -1 & 0 & 0 \\ 0 & -2 & 0 \\ 0 & 0 & -3 \end{bmatrix}$$

$$\tilde{\mathbf{B}} = \mathbf{V}^{-1}\mathbf{B} = \begin{bmatrix} 3 & 2.5 & 0.5 \\ -3 & -4 & -1 \\ 1 & 1.5 & 0.5 \end{bmatrix} \begin{bmatrix} 0 \\ 0 \\ 2 \end{bmatrix} = \begin{bmatrix} 1 \\ -2 \\ 1 \end{bmatrix}$$

$$\tilde{\mathbf{C}} = \mathbf{CV} = \begin{bmatrix} 1 & 0 & 0 \end{bmatrix} \begin{bmatrix} 1 & 1 & 1 \\ -1 & -2 & -3 \\ 1 & 4 & 9 \end{bmatrix} = \begin{bmatrix} 1 & 1 & 1 \end{bmatrix}.$$

Now, the DCF of the state model can be given as

$$\begin{bmatrix} \dot{z}_1(t) \\ \dot{z}_2(t) \\ \dot{z}_3(t) \end{bmatrix} = \begin{bmatrix} -1 & 0 & 0 \\ 0 & -2 & 0 \\ 0 & 0 & -3 \end{bmatrix} \begin{bmatrix} z_1(t) \\ z_2(t) \\ z_3(t) \end{bmatrix} + \begin{bmatrix} 1 \\ -2 \\ 1 \end{bmatrix} u(t)$$

$$y(t) = \begin{bmatrix} 1 & 1 & 1 \end{bmatrix} \begin{bmatrix} z_1(t) \\ z_2(t) \\ z_3(t) \end{bmatrix}$$ **Ans.**

Example 1.14. *A system is characterized by the transfer function*

$$\frac{Y(s)}{U(s)} = \frac{2}{s^3 + 6s^2 + 11s + 6}.$$

Represent the state model in Diagonal Canonical Form (DCF).

Solution. $$\frac{Y(s)}{U(s)} = \frac{2}{s^3 + 6s^2 + 11s + 6}$$

or, $$\frac{Y(s)}{U(s)} = \frac{2}{(s+1)(s+2)(s+3)}$$

or, $$Y(s) = \left[\frac{1}{s+1} - \frac{2}{s+2} + \frac{1}{s+3} \right] U(s). \qquad (1.109)$$

Define state-variables as follows

$$X_1(s) = \frac{1}{s+1}U(s)$$

$$X_2(s) = \frac{-2}{s+2}U(s)$$

$$X_3(s) = \frac{1}{s+3}U(s)$$

Then clearly,

$$sX_1(s) = -X_1(s) + U(s)$$
$$sX_2(s) = -2X_2(s) - 2U(s)$$
$$sX_3(s) = -3X_3(s) + U(s)$$

Taking the inverse Laplace transforms of the previous equations, assuming all initial conditions to be zero, we get

$$\dot{x}_1(t) = -x_1(t) + u(t)$$
$$\dot{x}_2(t) = -2x_2(t) - 2u(t)$$
$$\dot{x}_3(t) = -3x_3(t) + u(t)$$

or, in matrix form,

$$\begin{bmatrix} \dot{x}_1(t) \\ \dot{x}_2(t) \\ \dot{x}_3(t) \end{bmatrix} = \begin{bmatrix} -1 & 0 & 0 \\ 0 & -2 & 0 \\ 0 & 0 & -3 \end{bmatrix} \begin{bmatrix} x_1(t) \\ x_2(t) \\ x_3(t) \end{bmatrix} + \begin{bmatrix} 1 \\ -2 \\ 1 \end{bmatrix} u(t). \qquad (1.110\ A)$$

Now, in terms of state-variables, Equation (1.109) can be rewritten as

$$Y(s) = X_1(s) + X_2(s) + X_3(s).$$

Taking the inverse Laplace transform, we obtain

$$y(t) = x_1(t) + x_2(t) + x_3(t)$$

or,   in matrix form,

$$y(t) = \begin{bmatrix} 1 & 1 & 1 \end{bmatrix} \begin{bmatrix} x_1(t) \\ x_2(t) \\ x_3(t) \end{bmatrix}.$$ (1.110 B)

Equations (1.110) give the desired DCF of the state model.   **Ans.**

VARIANT:
The DCF can also be obtained using a similarity transformation. For this, convert the transfer function into controllable canonical form as follows

$$\frac{Y(s)}{U(s)} = \frac{2}{s^2 + 6s^2 + 11s + 6}$$

or, $$s^3 Y(s) = 2U(s) - 6s^2 Y(s) - 11sY(s) - 6Y(s).$$ (1.111)

Define state-variables as

$$X_1(s) = Y(s)$$
$$X_2(s) = sY(s)$$
$$X_3(s) = s^2 Y(s)$$

Then clearly,

$$\left. \begin{array}{l} sX_1(s) = sY(s) = X_2(s) \\ sX_2(s) = s^2 Y(s) = X_3(s) \\ sX_3(s) = s^3 Y(s) = 2U(s) - 6s^2 Y(s) - 11sY(s) - 6Y(s) \\ \qquad\qquad = 2U(s) - 6X_3(s) - 11X_2(s) - 6X_1(s) \end{array} \right]$$ (1.112)

Taking the inverse Laplace transforms of the set of equations (1.112); assuming all initial conditions to be zero, we get

$$\dot{x}_1(t) = x_2(t)$$
$$\dot{x}_2(t) = x_3(t)$$
$$\dot{x}_3(t) = 2u(t) - 6x_3(t) - 11x_2(t) - 6x_1(t)$$

or,   in matrix form,

$$\begin{bmatrix} \dot{x}_1(t) \\ \dot{x}_2(t) \\ \dot{x}_3(t) \end{bmatrix} = \begin{bmatrix} 0 & 1 & 0 \\ 0 & 0 & 1 \\ -6 & -11 & -6 \end{bmatrix} \begin{bmatrix} x_1(t) \\ x_2(t) \\ x_3(t) \end{bmatrix} + \begin{bmatrix} 0 \\ 0 \\ 2 \end{bmatrix} u(t). \qquad (1.113\ A)$$

Also, $Y(s) = X_1(s).$

Taking the inverse Laplace transform, we get

$$y(t) = x_1(t)$$

or, $$y(t) = \begin{bmatrix} 1 & 0 & 0 \end{bmatrix} \begin{bmatrix} x_1(t) \\ x_2(t) \\ x_3(t) \end{bmatrix}. \qquad (1.113\ B)$$

Equation (1.113) give the state-model in controllable canonical form.
Further proceeding as in Example 1.13 we will get the desired answer as

$$\begin{bmatrix} \dot{z}_1(t) \\ \dot{z}_2(t) \\ \dot{z}_3(t) \end{bmatrix} = \begin{bmatrix} -1 & 0 & 0 \\ 0 & -2 & 0 \\ 0 & 0 & -3 \end{bmatrix} \begin{bmatrix} z_1(t) \\ z_2(t) \\ z_3(t) \end{bmatrix} + \begin{bmatrix} 1 \\ -2 \\ 1 \end{bmatrix} u(t) \qquad (1.114\ A)$$

$$y(t) = \begin{bmatrix} 1 & 1 & 1 \end{bmatrix} \begin{bmatrix} z_1(t) \\ z_2(t) \\ z_3(t) \end{bmatrix}. \qquad (1.114\ B)$$

Equations (1.114) give the state model of the system in diagonal canonical form (DCF). **Ans.**

Example 1.15. *Consider the transfer Function*

$$\frac{Y(s)}{U(s)} = \frac{2s^2 + 6s + 7}{(s+1)^2(s+2)}.$$

Represent it in Jordan Canonical Form.

Solution. Notice that the system has repeated roots.

Now,
$$\frac{Y(s)}{U(s)} = \frac{2s^2 + 6s + 7}{(s+1)^2(s+2)}$$

or,
$$\frac{Y(s)}{U(s)} = \frac{3}{(s+1)^2} + \frac{(-1)}{(s+1)} + \frac{3}{(s+2)}$$

or,
$$Y(s) = \frac{3}{(s+1)^2}U(s) + \frac{(-1)}{(s+1)}U(s) + \frac{3}{(s+2)}U(s). \qquad (1.115)$$

Define state-variables as

$$X_1(s) = \frac{1}{(s+1)^2}U(s)$$

$$X_2(s) = \frac{1}{(s+1)}U(s)$$

$$X_3(s) = \frac{1}{(s+2)}U(s)$$

However, $X_1(s)$ and $X_2(s)$ are related as

$$\frac{X_1(s)}{X_2(s)} = \frac{1}{(s+1)}.$$

Then clearly,

$$sX_1(s) = -X_1(s) + X_2(s)$$
$$sX_2(s) = -X_2(s) + U(s)$$
$$sX_3(s) = -2X_3(s) + U(s)$$

and,
$$Y(s) = 3X_1(s) - X_2(s) + 3X_3(s)$$

Taking the inverse Laplace transforms, assuming initial conditions to be zero, we get

$$\left.\begin{aligned} \dot{x}_1(t) &= -x_1(t) + x_2(t) \\ \dot{x}_2(t) &= -x_2(t) + u(t) \\ \dot{x}_3(t) &= -2x_3(t) + u(t) \end{aligned}\right\} \text{State Equations}$$

and, $y(t) = 3x_1(t) - x_2(t) + 3x_3(t); \text{Output Equation}$

or,   in matrix form

Jordan Block

$$\begin{bmatrix} \dot{x}_1(t) \\ \dot{x}_2(t) \\ \dot{x}_3(t) \end{bmatrix} = \begin{bmatrix} -1 & 1 & 0 \\ 0 & -1 & 0 \\ 0 & 0 & -2 \end{bmatrix} \begin{bmatrix} x_1(t) \\ x_2(t) \\ x_3(t) \end{bmatrix} + \begin{bmatrix} 0 \\ 1 \\ 1 \end{bmatrix} u(t) \qquad (1.116\,A)$$

$$y(t) = \begin{bmatrix} 3 & -1 & 3 \end{bmatrix} \begin{bmatrix} x_1(t) \\ x_2(t) \\ x_3(t) \end{bmatrix}. \qquad (1.116\,B)$$

Equations (1.116) give the state model of the system in Jordan Canonical Form.   **Ans.**

1.15   CONCEPTS OF CONTROLLABILITY AND OBSERVABILITY

In order to decide whether a solution of control problems exists or not; there are two basic questions that need to be answered:

(a) Is it possible to transfer the system from any initial state to any desired state in a specified finite time by application of a suitable control force?
(b) Is it possible to determine the initial state of the system with the knowledge of the output vector for a finite time-interval?

Kalman conceptualized the answers to these basic questions into what is known as **controllability** and **observability.** The concepts of controllability and observability play an important role in control engineering. In fact, the conditions of controllability and observability may govern the existence of a complete solution to the control system design problems.

1.15.1 Complete State Controllability

Consider a single-input linear time-invariant system described by the state equation

$$\dot{\mathbf{x}}(t) = \mathbf{A}\mathbf{x}(t) + \mathbf{B}u(t \tag{1.117}$$

where,
$$\mathbf{x}(t) = \begin{bmatrix} x_1(t) \\ x_2(t) \\ \vdots \\ x_n(t) \end{bmatrix}_{n \times 1} ; \text{ State Vector}$$

$$\mathbf{A} = \begin{bmatrix} a_{11} & a_{12} & \cdots & a_{1n} \\ a_{21} & a_{22} & \cdots & A_{2n} \\ \vdots & \vdots & & \vdots \\ a_{n1} & a_{n2} & \cdots & a_{nn} \end{bmatrix}_{n \times n} ; \text{ System Matrix}$$

and,
$$\mathbf{B} = \begin{bmatrix} b_1 \\ b_2 \\ \vdots \\ b_n \end{bmatrix}_{n \times 1} ; \text{ Input Matrix.}$$

*"The system described by Equation (1.117) is said to be **completely state controllable**, if it is possible to transfer the system state from the initial state $\mathbf{x}(t_0)$ to the final state $\mathbf{x}(t_f)$ in a specified finite time-interval $t_0 \leq t \leq t_f$ by application of the control force $u(t)$."*

The concept of controllability involves the dependence of state-variables of the system on the control force. If any of the state-variables are independent of the control $u(t)$, there would be no way of driving this particular state-variable to the desired final state in a finite time-interval by means of a control effort. Therefore, this particular state is said to be uncontrollable and as long as there is at least one uncontrollable state. The system is said to be not completely controllable, or simply **uncontrollable**.

1.15.1.1 Gilbert Test for Complete State Controllability

Case 1: Let us assume that the eigenvalues λ_1, λ_2, ... λ_n of matrix **A** are all **distinct**.

The state Equation (1.117) can be transformed into the diagonal canonical form by means of a similarity transformation $\mathbf{x}(t) = \mathbf{M}\mathbf{z}(t)$ as

$$\dot{\mathbf{z}}(t) = \mathbf{\Lambda}\mathbf{z}(t) + \tilde{\mathbf{B}}u(t) \tag{1.118}$$

where, $\mathbf{\Lambda} = \mathbf{M}^{-1}\mathbf{A}\mathbf{M};$ A Diagonal Matrix

$\quad\quad\quad \tilde{\mathbf{B}} = \mathbf{M}^{-1}\mathbf{B}$

and **M** = Diagonalizing or Modal Matrix

Equation (1.118) may also be expressed as

$$\begin{bmatrix} \dot{z}_1(t) \\ \dot{z}_2(t) \\ \vdots \\ \dot{z}_n(t) \end{bmatrix}_{n\times 1} = \begin{bmatrix} \lambda_1 & 0 & \cdots & 0 \\ 0 & \lambda_2 & \cdots & 0 \\ \vdots & \vdots & & \vdots \\ 0 & 0 & \cdots & \lambda_n \end{bmatrix}_{n\times n} \begin{bmatrix} z_1(t) \\ z_2(t) \\ \vdots \\ z_n(t) \end{bmatrix}_{n\times 1} + \begin{bmatrix} \tilde{b}_1 \\ \tilde{b}_2 \\ \vdots \\ \tilde{b}_n \end{bmatrix}_{n\times 1} u(t) \tag{1.119}$$

or, in component form

$$\dot{z}_i(t) = \lambda_i z_i(t) + \tilde{b}_i u(t) \quad\quad\quad\quad (i = 1, 2, ..., n),$$

which has the solution

$$z_i(t) = e^{\lambda_i(t-t_0)} z_i(t_0) + \int_{t_0}^{t} e^{\lambda_i(t-\tau)}\, \tilde{b}_i u(\tau)a$$

or,

$$z_i(t) = e^{\lambda_i(t-t_0)} z_i(t_0) + e^{\lambda_i t}\int_{t_0}^{t} e^{-\lambda_i(\tau)}\, \tilde{b}_i u(\tau) d\tau.$$

The system described by Equation (1.119) is said to be **completely state controllable** if each state-variable $z_i(t)$ can be transferred from the initial state $z_i(t_0)$ to a final state $z_i(t_f)$ in a finite time-interval $t_0 \le t \le t_f$ ($i = 1, 2, ... n$).

In other words, the system is **completely state controllable** if it is possible to construct a control signal $u(t)$ such that

$$\frac{z_i(t_f) - e^{\lambda_i(t_f - t_0)} z_i(t_0)}{e^{\lambda_i t_f}} = \int_{t_0}^{t_f} e^{-\lambda_i \tau} \, \tilde{b}_i u(\tau) d\tau \qquad (i = 1, 2, \dots n).$$

Infact, there are numerous values of $u(\tau)$ that satisfy the preceeding condition provided $\tilde{b}_i \neq 0$, otherwise the **link** between the input and corresponding state-variable gets broken and hence it is no longer possible to control that particular state-variable.

It therefore follows that the **necessary condition** for **complete state controllability** is simply that the vector $\tilde{\mathbf{B}}$ must **not have any zero element**. If any element of this vector is zero, then the corresponding state-variable is not controllable and hence the system is not completely state controllable.

This result can be extended for **multi-input systems** also, where the control force $\mathbf{u}(t)$ is an $m \times 1$ vector. For the system described by

$$\dot{\mathbf{z}}(t) = \mathbf{\Lambda} \mathbf{z}(t) + \tilde{\mathbf{B}} \mathbf{u}(t)$$

where,
$$\tilde{\mathbf{B}} = \begin{bmatrix} \tilde{b}_{11} & \tilde{b}_{12} & \cdots & \tilde{b}_{1m} \\ \tilde{b}_{21} & \tilde{b}_{22} & \cdots & \tilde{b}_{2m} \\ \vdots & \vdots & & \vdots \\ \tilde{b}_{n1} & \tilde{b}_{n2} & \cdots & \tilde{b}_{nm} \end{bmatrix}_{n \times m}.$$

The **necessary condition** for **complete state controllability** is that the matrix $\tilde{\mathbf{B}}$ must have **no row with all zeros**. If any row of the matrix $\tilde{\mathbf{B}}$ is zero, it is not possible to influence the corresponding state-variable by the control forces and that particular state-variable is not controllable. Thus, the system is not completely state controllable.

Case 2: Let us assume that the matrix $\mathbf{A}$ has **multiple eigenvalues**. In this case the state equation

$$\dot{\mathbf{x}}(t) = \mathbf{A}\mathbf{x}(t) + \mathbf{B}\mathbf{u}(t)$$

can be transformed into the Jordan Canonical Form by means of a similarity transformation $\mathbf{x}(t) = \mathbf{S}\mathbf{z}(t)$ as

$$\dot{\mathbf{z}}(t) = \mathbf{J}\mathbf{z}(t) + \tilde{\mathbf{B}}\mathbf{u}(t) \tag{1.120}$$

where, $J = S^{-1}AS;$ A Jordan Matrix (Almost Diagonal)

$\tilde{B} = S^{-1}B.$

The **necessary condition** for **complete state controllability** is that the elements of any row of matrix $\tilde{B}$ that correspond to the **last row of each Jordan Block are not all zeros** and the elements of each row of $\tilde{B}$ that correspond to **distinct eigenvalues are not all zeros**.

Example 1.16. *Consider the following systems described by*

(a) $\begin{bmatrix} \dot{x}_1(t) \\ \dot{x}_2(t) \end{bmatrix} = \begin{bmatrix} -1 & 0 \\ 0 & -2 \end{bmatrix} \begin{bmatrix} x_1(t) \\ x_2(t) \end{bmatrix} + \begin{bmatrix} 3 \\ 5 \end{bmatrix} u(t)$

(b) $\begin{bmatrix} \dot{x}_1(t) \\ \dot{x}_2(t) \end{bmatrix} = \begin{bmatrix} -1 & 0 \\ 0 & -2 \end{bmatrix} \begin{bmatrix} x_1(t) \\ x_2(t) \end{bmatrix} + \begin{bmatrix} 3 \\ 0 \end{bmatrix} u(t)$

(c) $\begin{bmatrix} \dot{x}_1(t) \\ \dot{x}_2(t) \\ \dot{x}_3(t) \end{bmatrix} = \begin{bmatrix} -1 & 1 & 0 \\ 0 & -1 & 0 \\ 0 & 0 & -2 \end{bmatrix} \begin{bmatrix} x_1(t) \\ x_2(t) \\ x_3(t) \end{bmatrix} + \begin{bmatrix} 0 \\ 4 \\ 3 \end{bmatrix} u(t)$

(d) $\begin{bmatrix} \dot{x}_1(t) \\ \dot{x}_2(t) \\ \dot{x}_3(t) \end{bmatrix} = \begin{bmatrix} -1 & 1 & 0 \\ 0 & -1 & 0 \\ 0 & 0 & -2 \end{bmatrix} \begin{bmatrix} x_1(t) \\ x_2(t) \\ x_3(t) \end{bmatrix} + \begin{bmatrix} 4 & 2 \\ 0 & 0 \\ 3 & 0 \end{bmatrix} \begin{bmatrix} u_1(t) \\ u_2(t) \end{bmatrix}$

(e) $\begin{bmatrix} \dot{x}_1(t) \\ \dot{x}_2(t) \\ \dot{x}_3(t) \\ \dot{x}_4(t) \\ \dot{x}_5(t) \end{bmatrix} = \begin{bmatrix} -2 & 1 & 0 & 0 & 0 \\ 0 & -2 & 1 & 0 & 0 \\ 0 & 0 & -2 & 0 & 0 \\ 0 & 0 & 0 & -5 & 1 \\ 0 & 0 & 0 & 0 & -5 \end{bmatrix} \begin{bmatrix} x_1(t) \\ x_2(t) \\ x_3(t) \\ x_4(t) \\ x_5(t) \end{bmatrix} + \begin{bmatrix} 0 & 1 \\ 0 & 0 \\ 3 & 0 \\ 0 & 0 \\ 2 & 1 \end{bmatrix} \begin{bmatrix} u_1(t) \\ u_2(t) \end{bmatrix}$

and,   (f) $\begin{bmatrix} \dot{x}_1(t) \\ \dot{x}_2(t) \\ \dot{x}_3(t) \\ \dot{x}_4(t) \\ \dot{x}_5(t) \end{bmatrix} = \begin{bmatrix} -2 & 1 & 0 & 0 & 0 \\ 0 & -2 & 1 & 0 & 0 \\ 0 & 0 & -2 & 0 & 0 \\ \hline 0 & 0 & 0 & -5 & 1 \\ 0 & 0 & 0 & 0 & -5 \end{bmatrix} \begin{bmatrix} x_1(t) \\ x_2(t) \\ x_3(t) \\ x_4(t) \\ x_5(t) \end{bmatrix} + \begin{bmatrix} 4 \\ 2 \\ 1 \\ 3 \\ 0 \end{bmatrix} u(t).$

Solution. Applying the Gilbert Test, as discussed in Section 1.15.1.1, we conclude that the systems (a), (c), and (e) are completely state controllable. On the other hand, systems (b), (d), and (f) are not completely state controllable.   **Ans.**

Example 1.17. *Consider the system with the state equation*

$$\begin{bmatrix} \dot{x}_1(t) \\ \dot{x}_2(t) \\ \dot{x}_3(t) \end{bmatrix} = \begin{bmatrix} 0 & 1 & 0 \\ 0 & 0 & 1 \\ -6 & -11 & -6 \end{bmatrix} \begin{bmatrix} x_1(t) \\ x_2(t) \\ x_3(t) \end{bmatrix} + \begin{bmatrix} 0 \\ 0 \\ 2 \end{bmatrix} u(t). \qquad (1.121)$$

Determine whether the system is completely state controllable or not.
Solution. Compare Equation (1.121) with

$$\dot{\mathbf{x}}(t) = \mathbf{A}\mathbf{x}(t) + \mathbf{B}u(t).$$

We have, $\mathbf{A} = \begin{bmatrix} 0 & 1 & 0 \\ 0 & 0 & 1 \\ -6 & -11 & -6 \end{bmatrix}; \mathbf{B} = \begin{bmatrix} 0 \\ 0 \\ 2 \end{bmatrix}.$

The characteristic equation of matrix $\mathbf{A}$ is given by

$$|\lambda\mathbf{I} - \mathbf{A}| = 0.$$

Now, $\lambda\mathbf{I} - \mathbf{A} = \begin{bmatrix} \lambda & -1 & 0 \\ 0 & \lambda & -1 \\ 6 & 11 & \lambda+6 \end{bmatrix}$

$\therefore$ $\qquad$ $|\lambda \mathbf{I} - \mathbf{A}| = 0$

or, $\qquad$ $(\lambda + 1)(\lambda + 2)(\lambda + 3) = 0.$

Therefore, the eigenvalues of matrix $\mathbf{A}$ are

$$\lambda_1 = -1,$$
$$\lambda_2 = -2,$$
$$\lambda_3 = -3.$$

Matrix $\mathbf{A}$ is in Bush's form. Thus, we choose Vander Monde Matrix $\mathbf{V}$ as the modal matrix in order to transform Equation (1.121) in diagonal canonical form.

$$\mathbf{V} = \begin{bmatrix} 1 & 1 & 1 \\ \lambda_1 & \lambda_2 & \lambda_3 \\ \lambda_1^2 & \lambda_2^2 & \lambda_3^2 \end{bmatrix} = \begin{bmatrix} 1 & 1 & 1 \\ -1 & -2 & -3 \\ 1 & 4 & 9 \end{bmatrix}$$

$\therefore$ $\qquad$ $$\mathbf{V}^{-1} = \begin{bmatrix} 3 & 2.5 & 0.5 \\ -3 & -4 & -1 \\ 1 & 1.5 & 0.5 \end{bmatrix}$$

Using the similarity transformation $\mathbf{x}(t) = \mathbf{V}\mathbf{z}(t)$, the original state equation modifies to diagonal canonical form

$$\dot{\mathbf{z}}(t) = \mathbf{\Lambda}\mathbf{z}(t) + \tilde{\mathbf{B}}u(t)$$

where, $\qquad$ $\Lambda = V^{-1}AV$

and, $\qquad$ $\tilde{\mathbf{B}} = V^{-1}B.$

Now, $\qquad$ $\mathbf{\Lambda} = V^{-1}AV$

$$= \begin{bmatrix} 3 & 2.5 & 0.5 \\ -3 & -4 & -1 \\ 1 & 1.5 & 0.5 \end{bmatrix} \begin{bmatrix} 0 & 1 & 0 \\ 0 & 0 & 1 \\ -6 & -11 & -6 \end{bmatrix} \begin{bmatrix} 1 & 1 & 1 \\ -1 & -2 & -3 \\ 1 & 4 & 9 \end{bmatrix}$$

$$= \begin{bmatrix} -1 & 0 & 0 \\ 0 & -2 & 0 \\ 0 & 0 & -3 \end{bmatrix}$$

$$\tilde{\mathbf{B}} = \mathbf{V}^{-1}\mathbf{B} = \begin{bmatrix} 3 & 2.5 & 0.5 \\ -3 & -4 & -1 \\ 1 & 1.5 & 0.5 \end{bmatrix} \begin{bmatrix} 0 \\ 0 \\ 2 \end{bmatrix} = \begin{bmatrix} 1 \\ -2 \\ 1 \end{bmatrix}.$$

Now, the DCF of the state model can be given as

$$\begin{bmatrix} \dot{z}_1(t) \\ \dot{z}_2(t) \\ \dot{z}_3(t) \end{bmatrix} = \begin{bmatrix} -1 & 0 & 0 \\ 0 & -2 & 0 \\ 0 & 0 & -3 \end{bmatrix} \begin{bmatrix} z_1(t) \\ z_2(t) \\ z_3(t) \end{bmatrix} + \begin{bmatrix} 1 \\ -2 \\ 1 \end{bmatrix} u(t).$$

Applying the Gilbert Test we find that no element of $\tilde{\mathbf{B}}$ is zero, thus the system is completely state controllable.   **Ans.**

1.15.1.2 Kalman Test for Complete State Controllability

The Gilbert Test discussed in the previous section requires the system to be transformed into diagonal canonical or Jordan canonical form in order to test the complete state controllability. But the Kalman Test of controllability can be applied to any state model (diagonal canonical, Jordan canonical, or otherwise) and is stated here:

Consider a single-input linear time-invariant system

$$\dot{\mathbf{x}}(t) = \mathbf{A}\mathbf{x}(t) + \mathbf{B}u(t). \tag{1.122}$$

The system given by Equation (1.122) is completely state controllable **if, and only if**, the $n \times n$ composite matrix given by

$$\mathbf{Q}_c = [\mathbf{B} \vdots \mathbf{AB} \vdots \mathbf{A}^2\mathbf{B} \vdots \cdots\cdots \vdots \mathbf{A}^{n-1}\mathbf{B}]_{n \times n}$$

is of rank 'n.' This condition is also referred to as the pair $(\mathbf{A}, \mathbf{B})$ being **controllable**.

Now, consider a multi-input linear time-invariant system

$$\dot{\mathbf{x}}(t) = \mathbf{A}\mathbf{x}(t) + \mathbf{B}u(t). \tag{1.123}$$

Then, the system given by Equation (1.123) is completely state controllable **if, and only if**, the $n \times nm$ composite matrix given by

$$\mathbf{Q}_c = [\mathbf{B} \vdots \mathbf{AB} \vdots \mathbf{A}^2\mathbf{B} \vdots \cdots\cdots \vdots \mathbf{A}^{n-1} \mathbf{B}]_{n \times m}$$

is of **rank 'n.'**

$\mathbf{Q}_c$ is called the **Controllability Matrix**.

Example 1.18. *Consider the system with the state equation*

$$\begin{bmatrix} \dot{x}_1(t) \\ \dot{x}_2(t) \\ \dot{x}_3(t) \end{bmatrix} = \begin{bmatrix} 0 & 1 & 0 \\ 0 & 0 & 1 \\ -6 & -11 & -6 \end{bmatrix} \begin{bmatrix} x_1(t) \\ x_2(t) \\ x_3(t) \end{bmatrix} + \begin{bmatrix} 0 \\ 0 \\ 1 \end{bmatrix} u(t). \qquad (1.124)$$

Determine whether the system is completely state controllable or not.

Solution. Compare Equation (1.124) with

$$\dot{\mathbf{x}}(t) = \mathbf{Ax}(t) + \mathbf{B}u(t).$$

We have $\quad \mathbf{A} = \begin{bmatrix} 0 & 1 & 0 \\ 0 & 0 & 1 \\ -6 & -11 & -6 \end{bmatrix}; \mathbf{B} = \begin{bmatrix} 0 \\ 0 \\ 1 \end{bmatrix}.$

Then,

$$\mathbf{AB} = \begin{bmatrix} 0 & 1 & 0 \\ 0 & 0 & 1 \\ -6 & -11 & -6 \end{bmatrix} \begin{bmatrix} 0 \\ 0 \\ 1 \end{bmatrix} = \begin{bmatrix} 0 \\ 1 \\ -6 \end{bmatrix}$$

and, $\quad \mathbf{A}^2\mathbf{B} = \mathbf{A}(\mathbf{AB})$

$$= \begin{bmatrix} 0 & 1 & 0 \\ 0 & 0 & 1 \\ -6 & -11 & -6 \end{bmatrix} \begin{bmatrix} 0 \\ 1 \\ -6 \end{bmatrix} = \begin{bmatrix} 1 \\ -6 \\ 25 \end{bmatrix}.$$

Now, the composite matrix $\mathbf{Q}_c$ is given by

$$\mathbf{Q}_c = [\mathbf{B}\,\vdots\,\mathbf{AB}\,\vdots\,\mathbf{A}^2\mathbf{B}]$$

$$= \begin{bmatrix} 0 & 0 & 1 \\ 0 & 1 & -6 \\ 1 & -6 & 25 \end{bmatrix}.$$

It is easily seen that   det $\mathbf{Q}_c \neq 0$ i.e.,

$$\text{Rank } \mathbf{Q}_c = 3 = n.$$

The system is, therefore, completely state controllable (as per the Kalman Test).   **Ans.**

1.15.2 Complete Output Controllability

We sometimes may want to control the output rather than the state of the system. Complete state controllability is **neither necessary nor sufficient** for controlling the output of the system. For this reason, it is desirable to define separately the complete output controllability.

Consider a multi-input-multi-output linear time-invariant system

$$\dot{\mathbf{x}}(t) = \mathbf{A}\mathbf{x}(t) + \mathbf{B}\mathbf{u}(t) \tag{1.125 A}$$

$$\mathbf{y}(t) = \mathbf{C}\mathbf{x}(t) \tag{1.125 B}$$

where, $\mathbf{x}(t) = \begin{bmatrix} \dot{x}_1(t) \\ \dot{x}_2(t) \\ \vdots \\ \dot{x}_n(t) \end{bmatrix}_{n \times 1}$; State Vector

$$\mathbf{u}(t) = \begin{bmatrix} u_1(t) \\ u_2(t) \\ \vdots \\ u_m(t) \end{bmatrix}_{m \times 1} ; \quad \text{Input Vector}$$

$$\mathbf{A} = \begin{bmatrix} a_{11} & a_{12} & \cdots\cdots & a_{1n} \\ a_{21} & a_{22} & \cdots\cdots & a_{2n} \\ \vdots & \vdots & \cdots\cdots & \vdots \\ a_{n1} & a_{n2} & \cdots\cdots & a_{nn} \end{bmatrix}_{n \times n} ; \text{ System Matrix}$$

$$\mathbf{B} = \begin{bmatrix} b_{11} & b_{12} & \cdots\cdots & b_{1m} \\ b_{21} & b_{22} & \cdots\cdots & b_{2m} \\ \vdots & \vdots & \cdots\cdots & \vdots \\ b_{n1} & b_{n2} & \cdots\cdots & b_{nm} \end{bmatrix}_{n \times m} ; \text{Input Matrix}$$

and,

$$\mathbf{C} = \begin{bmatrix} c_{11} & c_{12} & \cdots\cdots & c_{1n} \\ c_{21} & c_{22} & \cdots\cdots & c_{2n} \\ \vdots & \vdots & \cdots\cdots & \vdots \\ c_{p1} & c_{p2} & \cdots\cdots & c_{pn} \end{bmatrix}_{p \times n} ; \text{ Output Matrix.}$$

Then, the Kalman Test of complete output controllability states that:

The system given by Equations (1.125) is completely output controllable **if, and only if**, the $p \times nm$ composite matrix given by

$$\mathbf{P}_c = [\mathbf{CB} \vdots \mathbf{CAB} \vdots \mathbf{CA}^2\mathbf{B} \vdots \cdots\cdots \vdots \mathbf{CA}^{n-1}\mathbf{B}]_{p \times nm}$$

is of **rank** 'p.'

1.15.3 Complete Observability

Consider a single-input-single-output linear time-invariant system described by

$$\dot{\mathbf{x}}(t) = \mathbf{A}\mathbf{x}(t) + \mathbf{B}u(t) \qquad\qquad (1.126\ A)$$

$$\mathbf{y}(t) = \mathbf{C}\mathbf{x}(t) \qquad\qquad (1.126\ B)$$

where,

$$\mathbf{x}(t) = \begin{bmatrix} x_1(t) \\ x_2(t) \\ \vdots \\ x_n(t) \end{bmatrix}_{n \times 1} ; \qquad \text{State Vector}$$

$$\mathbf{A} = \begin{bmatrix} a_{11} & a_{12} & ...... & a_{1n} \\ a_{21} & a_{22} & ...... & a_{2n} \\ \vdots & \vdots & ...... & \vdots \\ a_{n1} & a_{n2} & ...... & a_{nn} \end{bmatrix}_{n\times n} ; \qquad \text{System Matrix}$$

$$\mathbf{B} = \begin{bmatrix} b_1 \\ b_2 \\ \vdots \\ b_n \end{bmatrix}_{n\times 1} ; \qquad \text{Input Matrix}$$

and,
$$\mathbf{C} = [c_1 \ \ c_2 \ .... \ c_n]_{1\times n} ; \qquad \text{Output Matrix.}$$

"The system described by Equations (1.126) is said to be **completely observable,** *if every initial state* $\mathbf{x}(t_0)$ *can be completely determined with the knowledge of the output* $y(t)$ *over a finite time-interval* $t_0 \le t \le t_f$*".*

Essentially, a system is completely observable if every state-variable of the system affects the output. If any of the states cannot be observed from the measurement of the output, the state is said to be unobservable and the system is not completely observable, or is simply **unobservable**.

1.15.3.1 Gilbert Test for Complete Observability

Case 1: Let us assume that the eigenvalues $\lambda_1, \lambda_2, \ldots\ldots \lambda_n$ of matrix **A** are all **distinct**.

The state model (1.126) can be transformed into the diagonal canonical form by means of a similarity transformation $\mathbf{x}(t) = \mathbf{Mz}(t)$ as

$$\dot{\mathbf{z}}(t) = \Lambda\mathbf{z}(t) + \tilde{\mathbf{B}}u(t) \qquad (1.127\ A)$$

$$y(t) = \tilde{\mathbf{C}}\mathbf{z}(t) \qquad (1.127\ B)$$

where, $\Lambda = \mathbf{M}^{-1}\mathbf{AM}$; A Diagonal Matrix

$\tilde{\mathbf{B}} = \mathbf{M}^{-1}\mathbf{B}$

$\tilde{\mathbf{C}} = \mathbf{CM}$

and,   M = Diagonalizing or Modal Matrix.

Equation $(1.127\ B)$ can also be expressed as

$$y(t) = [\tilde{c}_1 \quad \tilde{c}_2 \quad \tilde{c}_3 \quad \ldots\ldots \quad \tilde{c}_n]_{1 \times n} \begin{bmatrix} z_1(t) \\ z_2(t) \\ \vdots \\ z_n(t) \end{bmatrix}_{n \times 1}$$

or, $\qquad y(t) = \tilde{c}_1 z_1(t) + \tilde{c}_2 z_2(t) + \ldots\ldots + \tilde{c}_n z_n(t).$ $\qquad (1.127\ C)$

Since diagonalization **decouples** the states, no state contains any information regarding any other state. Therefore, each state must be **independently observable** for complete observability. It follows that for a state to be observed through the output $y(t)$, its corresponding coefficient in Equation $(1.127\ C)$ must be nonzero.

If any particular $\tilde{c}_i$ is zero, the corresponding $z_i(t)$ can have any value without its effect showing up in the output $y(t)$. Thus, the **necessary condition** for **complete observability** is simply that the vector $\tilde{\mathbf{C}}$ must **not have any zero element**.

This result can be extended for **multi-input-multi-output systems** where the output vector $y(t)$, after transformation in diagonal canonical form, is given by

$$\mathbf{y}(t) = \tilde{\mathbf{C}}\mathbf{z}(t)$$

where, $\qquad \tilde{\mathbf{C}} = \begin{bmatrix} \tilde{c}_{11} & \tilde{c}_{12} & \ldots\ldots & \tilde{c}_{1n} \\ \tilde{c}_{21} & \tilde{c}_{22} & \ldots\ldots & \tilde{c}_{2n} \\ \vdots & \vdots & \ldots\ldots & \vdots \\ \tilde{c}_{p1} & \tilde{c}_{p2} & \ldots\ldots & \tilde{c}_{pm} \end{bmatrix}_{p \times n}.$

The **necessary condition** for **complete observability** is that the matrix $\tilde{\mathbf{C}}$ must have **no column with all zeros**.

Case 2: Let us assume that the matrix $\mathbf{A}$ has **multiple eigenvalues**. In this case the state model

$$\dot{\mathbf{x}}(t) = \mathbf{A}\mathbf{x}(t) + \mathbf{B}\mathbf{u}(t)$$
$$\mathbf{y}(t) = \mathbf{C}\mathbf{x}(t)$$

can be transformed into the Jordan canonical form by means of a similarity transformation $\mathbf{x}(t) = \mathbf{Sz}(t)$ as

$$\dot{\mathbf{z}}(t) = \mathbf{Jz}(t) + \tilde{\mathbf{B}}\mathbf{u}(t) \qquad (1.128\ A)$$

$$\mathbf{y}(t) = \tilde{\mathbf{C}}\mathbf{z}(t) \qquad (1.128\ B)$$

where, $\mathbf{J} = \mathbf{S}^{-1}\mathbf{AS}$;   A Jordan Matrix (Almost Diagonal)

$\tilde{\mathbf{B}} = \mathbf{S}^{-1}\mathbf{B}$

and, $\tilde{\mathbf{C}} = \mathbf{CS}$.

The **necessary condition** for **complete observability** is that the elements of any column of matrix $\tilde{\mathbf{C}}$ that correspond to the **first row of each Jordan Block are not all zeros** and the elements of each column of $\tilde{\mathbf{C}}$ that correspond to **distinct eigenvalues are not all zeros**.

Example 1.19. *Consider the following systems described by*

(a) $$\begin{bmatrix} \dot{x}_1(t) \\ \dot{x}_2(t) \end{bmatrix} = \begin{bmatrix} -1 & 0 \\ 0 & -2 \end{bmatrix} \begin{bmatrix} x_1(t) \\ x_2(t) \end{bmatrix}$$

$$y(t) = \begin{bmatrix} 1 & 3 \end{bmatrix} \begin{bmatrix} x_1(t) \\ x_2(t) \end{bmatrix}$$

(b) $$\begin{bmatrix} \dot{x}_1(t) \\ \dot{x}_2(t) \end{bmatrix} = \begin{bmatrix} -1 & 0 \\ 0 & -2 \end{bmatrix} \begin{bmatrix} x_1(t) \\ x_2(t) \end{bmatrix}$$

$$y(t) = \begin{bmatrix} 0 & 1 \end{bmatrix} \begin{bmatrix} x_1(t) \\ x_2(t) \end{bmatrix}$$

(c) $$\begin{bmatrix} \dot{x}_1(t) \\ \dot{x}_2(t) \\ \dot{x}_3(t) \end{bmatrix} = \begin{bmatrix} 2 & 1 & 0 \\ 0 & 2 & 1 \\ 0 & 0 & 2 \end{bmatrix} \begin{bmatrix} x_1(t) \\ x_2(t) \\ x_3(t) \end{bmatrix}$$

$$\begin{bmatrix} y_1(t) \\ y_2(t) \end{bmatrix} = \begin{bmatrix} 3 & 0 & 0 \\ 4 & 0 & 0 \end{bmatrix} \begin{bmatrix} x_1(t) \\ x_2(t) \\ x_3(t) \end{bmatrix}$$

(d) $$\begin{bmatrix} \dot{x}_1(t) \\ \dot{x}_2(t) \\ \dot{x}_3(t) \end{bmatrix} = \begin{bmatrix} 2 & 1 & 0 \\ 0 & 2 & 1 \\ 0 & 0 & 2 \end{bmatrix} \begin{bmatrix} x_1(t) \\ x_2(t) \\ x_3(t) \end{bmatrix}$$

$$\begin{bmatrix} y_1(t) \\ y_2(t) \end{bmatrix} = \begin{bmatrix} 0 & 1 & 3 \\ 0 & 2 & 4 \end{bmatrix} \begin{bmatrix} x_1(t) \\ x_2(t) \\ x_3(t) \end{bmatrix}$$

$$(e)\quad \begin{bmatrix} \dot{x}_1(t) \\ \dot{x}_2(t) \\ \dot{x}_3(t) \\ \dot{x}_4(t) \\ \dot{x}_5(t) \end{bmatrix} = \begin{bmatrix} 2 & 1 & 0 & 0 & 0 \\ 0 & 2 & 1 & 0 & 0 \\ 0 & 0 & 2 & 0 & 0 \\ 0 & 0 & 0 & -3 & 1 \\ 0 & 0 & 0 & 0 & -3 \end{bmatrix} \begin{bmatrix} x_1(t) \\ x_2(t) \\ x_3(t) \\ x_4(t) \\ x_5(t) \end{bmatrix} \qquad (f)\quad \begin{bmatrix} \dot{x}_1(t) \\ \dot{x}_2(t) \\ \dot{x}_3(t) \\ \dot{x}_4(t) \\ \dot{x}_5(t) \end{bmatrix} = \begin{bmatrix} 2 & 1 & 0 & 0 & 0 \\ 0 & 2 & 1 & 0 & 0 \\ 0 & 0 & 2 & 0 & 0 \\ 0 & 0 & 0 & -3 & 1 \\ 0 & 0 & 0 & 0 & -3 \end{bmatrix} \begin{bmatrix} x_1(t) \\ x_2(t) \\ x_3(t) \\ x_4(t) \\ x_5(t) \end{bmatrix}$$

$$\begin{bmatrix} y_1(t) \\ y_2(t) \end{bmatrix} = \begin{bmatrix} 1 & 1 & 1 & 0 & 0 \\ 0 & 1 & 1 & 1 & 0 \end{bmatrix} \begin{bmatrix} x_1(t) \\ x_2(t) \\ x_3(t) \\ x_4(t) \\ x_5(t) \end{bmatrix} \qquad \begin{bmatrix} y_1(t) \\ y_2(t) \end{bmatrix} = \begin{bmatrix} 1 & 1 & 1 & 0 & 0 \\ 0 & 1 & 1 & 0 & 0 \end{bmatrix} \begin{bmatrix} x_1(t) \\ x_2(t) \\ x_3(t) \\ x_4(t) \\ x_5(t) \end{bmatrix}.$$

Determine whether all the above systems are completely observable or not.

Solution. Here, the columns of $\widetilde{\mathbf{C}}$ have been encircled by dashed lines that correspond to the first row of each Jordan Block.

Applying the Gilbert Test, we conclude that the systems (a), (c), and (e) are completely observable. On the other hand, systems (b), (d), and (e) are not completely observable.   **Ans.**

Example 1.20. *Consider the system described as*

$$\begin{bmatrix} \dot{x}_1(t) \\ \dot{x}_2(t) \\ \dot{x}_3(t) \end{bmatrix} = \begin{bmatrix} 0 & 1 & 0 \\ 0 & 0 & 1 \\ 0 & -2 & -3 \end{bmatrix} \begin{bmatrix} x_1(t) \\ x_2(t) \\ x_3(t) \end{bmatrix} + \begin{bmatrix} 0 \\ 0 \\ 1 \end{bmatrix} u(t)$$

$$y(t) = \begin{bmatrix} 3 & 4 & 1 \end{bmatrix} \begin{bmatrix} x_1(t) \\ x_2(t) \\ x_3(t) \end{bmatrix}.$$

Determine whether the system is observable or not.

Solution. Here, $\mathbf{A} = \begin{bmatrix} 0 & 1 & 0 \\ 0 & 0 & 1 \\ 0 & -2 & -3 \end{bmatrix}$; $\mathbf{B} = \begin{bmatrix} 0 \\ 0 \\ 1 \end{bmatrix}$ and, $\mathbf{C} = \begin{bmatrix} 3 & 4 & 1 \end{bmatrix}$.

The characteristic equation of matrix $\mathbf{A}$ is given by

$$|\lambda\mathbf{I} - \mathbf{A}| = 0$$

or,

$$\begin{vmatrix} \lambda & -1 & 0 \\ 0 & \lambda & -1 \\ 0 & 2 & \lambda+3 \end{vmatrix} = 0$$

or,

$$\lambda(\lambda+1)(\lambda+2) = 0.$$

Therefore, the eigenvalues of matrix $\mathbf{A}$ are

$$\lambda_1 = 0,$$
$$\lambda_2 = -1,$$
$$\lambda_3 = -2.$$

As the matrix $\mathbf{A}$ is in Bush's form, we choose Vander Monde Matrix $\mathbf{V}$ as the modal matrix, given by

$$\mathbf{V} = \begin{bmatrix} 1 & 1 & 1 \\ \lambda_1 & \lambda_2 & \lambda_3 \\ \lambda_1^2 & \lambda_2^2 & \lambda_3^2 \end{bmatrix} = \begin{bmatrix} 1 & 1 & 1 \\ 0 & -1 & -2 \\ 0 & 1 & 4 \end{bmatrix}.$$

Under similarity transformation $\mathbf{x}(t) = \mathbf{V}\mathbf{z}(t)$, the output is given by

$$y(t) = \mathbf{C}\mathbf{V}\mathbf{z}(t) = \mathbf{C}\mathbf{z}(t)$$

where,

$$\tilde{\mathbf{C}} = \begin{bmatrix} 3 & 4 & 1 \end{bmatrix}\begin{bmatrix} 1 & 1 & 1 \\ 0 & -1 & -2 \\ 0 & 1 & 4 \end{bmatrix}$$

or,

$$\tilde{\mathbf{C}} = \begin{bmatrix} 3 & 0 & -1 \end{bmatrix}.$$

Applying the Gilbert Test, we find that one element of matrix $\tilde{\mathbf{C}}$ is zero. Thus, the system is not completely observable. **Ans.**

1.15.3.2 Kalman Test for Complete Observability

Consider a single-input-single-output linear time-invariant system

$$\dot{\mathbf{x}}(t) = \mathbf{A}\mathbf{x}(t) + \mathbf{B}u(t) \tag{1.129 A}$$
$$y(t) = \mathbf{C}\mathbf{x}(t). \tag{1.129 B}$$

The system described by Equations (1.129) is completely observable **if, and only if**, the $n \times n$ composite matrix given by

$$\mathbf{Q}_0 = [\mathbf{C}^{\mathrm{T}} \vdots \mathbf{A}^{\mathrm{T}}\mathbf{C}^{\mathrm{T}} \vdots (\mathbf{A}^{\mathrm{T}})^2 \mathbf{C}^{\mathrm{T}} \vdots \cdots \vdots (\mathbf{A}^{\mathrm{T}})^{n-1}\mathbf{C}^{\mathrm{T}}]_{n \times n}$$

is of **rank 'n.'**

Now, consider a multi-input-multi-output linear time-invariant system

$$\dot{\mathbf{x}}(t) = \mathbf{A}\mathbf{x}(t) + \mathbf{B}u(t) \tag{1.130 A}$$
$$\mathbf{y}(t) = \mathbf{C}\mathbf{x}(t). \tag{1.130 B}$$

Then, the system described by Equations (1.130) is completely observable **if, and only if**, the $n \times np$ composite matrix given by

$$\mathbf{Q}_0 = [\mathbf{C}^{\mathrm{T}} \vdots \mathbf{A}^{\mathrm{T}}\mathbf{C}^{\mathrm{T}} \vdots (\mathbf{A}^{\mathrm{T}})^2 \mathbf{C}^{\mathrm{T}} \vdots \cdots \vdots (\mathbf{A}^{\mathrm{T}})^{n-1}\mathbf{C}^{\mathrm{T}}]_{n \times np}$$

is of **rank 'n.'**

This condition is also referred to as the pair $(\mathbf{A}, \mathbf{C})$ being **observable**. $\mathbf{Q}_0$ is referred to as the **observability matrix**.

Example 1.21. *Examine the observability of the system given in example 1.20 by the Kalman Test.*

Solution. Here, $\mathbf{A} = \begin{bmatrix} 0 & 1 & 0 \\ 0 & 0 & 1 \\ 0 & -2 & -3 \end{bmatrix}$; $\mathbf{B} = \begin{bmatrix} 0 \\ 0 \\ 1 \end{bmatrix}$ and $\mathbf{C} = [3 \quad 4 \quad 1]$

We have
$$\mathbf{C}^{\mathrm{T}} = \begin{bmatrix} 3 \\ 4 \\ 1 \end{bmatrix}$$

$$\mathbf{A}^{\mathrm{T}}\mathbf{C}^{\mathrm{T}} = \begin{bmatrix} 0 & 0 & 0 \\ 1 & 0 & -2 \\ 0 & 1 & -3 \end{bmatrix} \begin{bmatrix} 3 \\ 4 \\ 1 \end{bmatrix} = \begin{bmatrix} 0 \\ 1 \\ 1 \end{bmatrix}$$

$$(\mathbf{A}^{\mathrm{T}})^2\mathbf{C}^{\mathrm{T}} = \begin{bmatrix} 0 & 0 & 0 \\ 1 & 0 & -2 \\ 0 & 1 & -3 \end{bmatrix} \begin{bmatrix} 0 \\ 1 \\ 1 \end{bmatrix} = \begin{bmatrix} 0 \\ -2 \\ -2 \end{bmatrix}.$$

Therefore, the composite matrix $\mathbf{Q}_0$ is given by

$$\mathbf{Q}_0 = [\mathbf{C}^{\mathrm{T}} \vdots \mathbf{A}^{\mathrm{T}}\mathbf{C}^{\mathrm{T}} \vdots (\mathbf{A}^{\mathrm{T}})^2\mathbf{C}^{\mathrm{T}}]$$

or,
$$\mathbf{Q}_0 = \begin{bmatrix} 3 & 0 & 0 \\ 4 & 1 & -2 \\ 1 & 1 & -2 \end{bmatrix}.$$

Since
$$\begin{vmatrix} 3 & 0 \\ 4 & 1 \end{vmatrix} \neq 0 \text{ and } \begin{vmatrix} 3 & 0 & 0 \\ 4 & 1 & -2 \\ 1 & 1 & -2 \end{vmatrix} = 0$$

∴   Rank $\mathbf{Q}_0 = 2$ while $n = 3$.

Hence, one of the state-variables is unobservable i.e., the system is not completely observable.   **Ans.**

1.15.4 Relationship between Controllability and Observability (Principle of Duality)

We shall now discuss the relationship between controllability and observability by introducing the principle of duality, due to Kalman to clarify the **apparent analogies** between controllability and observability.

The principle of duality states that

(a) The pair (**A, B**) is **controllable** if, and only if, the pair (**A**^{T}, **B**^{T}) is **observable**.

(b) The pair ($\mathbf{A}$, $\mathbf{C}$) is observable if, and only if, the pair ($\mathbf{A^T}$, $\mathbf{B^T}$) is **controllable**.

1.15.5 Effect of Pole-Zero Cancellation in Transfer Function on Controllability and Observability

Consider the transfer function

$$G(s) = \frac{Y(s)}{U(s)} = \frac{b_0 s^m + b_1 s^{m-1} + ... + b_{m-1} s + b_m}{s^n + a_1 s^{n-1} + ... + a_{n-1} s + a_n}$$

$$= \frac{\beta(s - \alpha_1)(s - \alpha_2)......(s - \alpha_m)}{(s - \lambda_1)(s - \lambda_2)......(s - \lambda_n)} \qquad (\beta \text{ is a constant})$$

$$= \sum_{i=1}^{n} \frac{K_i}{(s - \lambda_i)}$$

where K_i is the residue of poles at $s = \lambda_i$.

Assume that the transfer function has pole-zero cancellation i.e., the transfer function has an **identical pair** of pole and zero at $\alpha_i = \lambda_i$; thus, $K_i = 0$. Because of this cancellation, the system will either be state uncontrollable or unobservable, **depending on how the state-variables are defined.** For example, if state-variables are selected so as to get the state model of the form (1.94), then $K_i = 0$ will appear in output vector $\mathbf{C}$ and the corresponding state $x_i(t)$ is **shielded from observation.** On the other hand, if state-variables are selected so as to get the state model of the form (1.96), then $K_i = 0$ will **appear** in control vector $\mathbf{B}$ and the state $x_i(t)$ is **uncontrollable**.

If the transfer function does not have pole-zero cancellation, the system can always be represented by completely controllable and observable state models.

EXERCISES

1. Determine STM when A is given by,

$$\mathbf{A} = \begin{bmatrix} 0 & 1 & 0 \\ 0 & 0 & 1 \\ 1 & -3 & 3 \end{bmatrix}.$$

2. Given $\mathbf{A} = \begin{bmatrix} 0 & 1 \\ 0 & -2 \end{bmatrix}$, compute $e^{\mathbf{A}t}$.

3. Obtain the response $y(t)$ of the system, described by

$$\begin{bmatrix} \dot{x}_1 \\ \dot{x}_2 \end{bmatrix} = \begin{bmatrix} -1 & -0.5 \\ 1 & 0 \end{bmatrix} \begin{bmatrix} x_1 \\ x_2 \end{bmatrix} + \begin{bmatrix} 0.5 \\ 0 \end{bmatrix} u$$

$$y = \begin{bmatrix} 1 & 0 \end{bmatrix} \begin{bmatrix} x_1 \\ x_2 \end{bmatrix}$$

where $u(t)$ is the unit step occurring at $t = 0$ and $\mathbf{x}^T(0) = [0\ 0]$.

4. A system, described by

$$\dot{\mathbf{x}}(t) = \begin{bmatrix} -1 & 1 \\ 0 & -2 \end{bmatrix} \begin{bmatrix} x_1 \\ x_2 \end{bmatrix} + \begin{bmatrix} 1 & 0 & 1 \\ 0 & 1 & 1 \end{bmatrix} u(t)$$

$$y(t) = \begin{bmatrix} 1 & 2 \\ 1 & 0 \\ 1 & 1 \end{bmatrix} \mathbf{x}(t).$$

Obtain the transfer function of the system.

5. Consider a system $\dot{\mathbf{x}}(t) = \mathbf{A}\mathbf{x}(t)$. For this system with $\mathbf{x}(0) = \begin{bmatrix} 1 \\ -2 \end{bmatrix}$ the response is given by

$$\mathbf{x}(t) = \begin{bmatrix} e^{-2t} \\ -2e^{-2t} \end{bmatrix}.$$

Determine the system matrix $\mathbf{A}$ and the state transition matrix.

6. Consider the following matrix

$$A = \begin{bmatrix} 0 & 1 \\ -6 & -5 \end{bmatrix}.$$

Find the STM, and also determine x(t), given that

$$x(0) = \begin{bmatrix} 1 \\ 0 \end{bmatrix}.$$

7. Given,

$$A_1 = \begin{bmatrix} \sigma & 0 \\ 0 & \sigma \end{bmatrix}; \quad A_2 = \begin{bmatrix} 0 & \omega \\ -\omega & \sigma \end{bmatrix} \quad and \quad A = \begin{bmatrix} \sigma & \omega \\ -\omega & \sigma \end{bmatrix}$$

compute e^{At}.

[**Hint.** $e^{At} = e^{(A_1 + A_2)t} = e^{A_1 t} \cdot e^{A_2 t}$ provided $A_1 A_2 = A_2 A_1$]

8. Show that the STM for the state equation

$$\begin{bmatrix} \dot{x}_1 \\ \dot{x}_2 \end{bmatrix} = \begin{bmatrix} -3 & 1 \\ 0 & -1 \end{bmatrix} \begin{bmatrix} x_1 \\ x_2 \end{bmatrix} \quad is \quad \begin{bmatrix} e^{-3t} & \frac{1}{2}(e^{-t} - e^{-3t}) \\ 0 & -e^{-t} \end{bmatrix}.$$

9. A system is described by the following differential equation

$$\dddot{x} + 3\ddot{x} + 4\dot{x} + 4x = u_1 + 3u_2 + 4u_3$$

and the outputs are

$$y_1 = 4\dot{x} + 3u_1$$

$$y_2 = \ddot{x} + 4u_2 + u_3.$$

Represent the system in state-space.

10. A single-input-single-output system is given by

$$\dot{x}(t) = \begin{bmatrix} -1 & 0 & 0 \\ 0 & -2 & 0 \\ 0 & 0 & -3 \end{bmatrix} \begin{bmatrix} x_1(t) \\ x_2(t) \\ x_3(t) \end{bmatrix} + \begin{bmatrix} 1 \\ 1 \\ 0 \end{bmatrix} u(t)$$

$$y(t) = \begin{bmatrix} 1 & 0 & 2 \end{bmatrix} x(t).$$

Test for controllability and observability.

11. Discuss the controllability of the systems described by

(a) $\begin{bmatrix} \dot{x}_1 \\ \dot{x}_2 \end{bmatrix} = \begin{bmatrix} 1 & 1 \\ 2 & -1 \end{bmatrix} \begin{bmatrix} x_1 \\ x_2 \end{bmatrix} + \begin{bmatrix} 0 \\ 1 \end{bmatrix} [u]$ (b) $\begin{bmatrix} \dot{x}_1 \\ \dot{x}_2 \end{bmatrix} = \begin{bmatrix} 1 & 1 \\ 0 & -1 \end{bmatrix} \begin{bmatrix} x_1 \\ x_2 \end{bmatrix} + \begin{bmatrix} 0 \\ 1 \end{bmatrix} u.$

12. Write the state equations for the circuit shown in Figure 1.13.

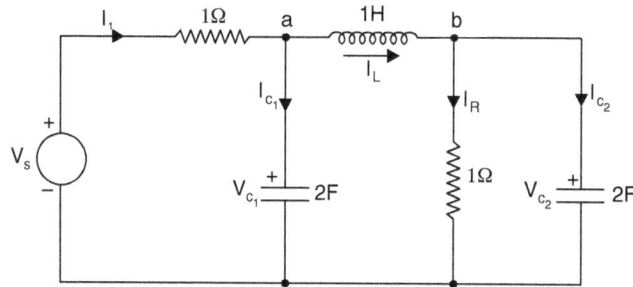

FIGURE 1.13

13. Consider the system described by

$$\begin{bmatrix} \dot{x}_1 \\ \dot{x}_2 \end{bmatrix} = \begin{bmatrix} 1 & 1 \\ -2 & -1 \end{bmatrix} \begin{bmatrix} x_1 \\ x_2 \end{bmatrix} + \begin{bmatrix} 0 \\ 1 \end{bmatrix} u$$

$$y = \begin{bmatrix} 1 & 0 \end{bmatrix} \begin{bmatrix} x_1 \\ x_2 \end{bmatrix}.$$

Is this system controllable and observable?

14. Consider the system described by the transfer function,

$$\frac{Y(s)}{U(s)} = \frac{s+3}{s^2 + 3s + 2}.$$

Obtain state-space representation in CCF, OCF, and in DCF.

15. Use diagonalization of matrix A to determine the time-response of the system:

$$\begin{bmatrix} \dot{x}_1 \\ \dot{x}_2 \end{bmatrix} = \begin{bmatrix} 1 & 1 \\ -6 & -5 \end{bmatrix} \begin{bmatrix} x_1 \\ x_2 \end{bmatrix} + \begin{bmatrix} 1 \\ 0 \end{bmatrix} u$$

and

$$y = [6 \quad 1] \begin{bmatrix} x_1 \\ x_2 \end{bmatrix}$$

given that

$$\begin{bmatrix} x_1(0) \\ x_2(0) \end{bmatrix} = \begin{bmatrix} 1 \\ 0 \end{bmatrix}.$$

REFERENCES

(1) Benjamin C. KuO, *Automatic Control Systems*, Prentice-Hall, Englewood Cliffs, New Jersey, 7th Edition, 1995.

(2) B.S. Manke, *Linear Control Systems*, Khanna Publishers, Delhi, 6th Edition, 2003.

(3) D. Roy Choudhury, *Modern Control Engineering*, Prentice-Hall of India Pvt. Ltd., New Delhi, 2005.

(4) Francis H. Raven, *Automatic Control Engineering*, McGraw Hill, Inc., Singapore, 4th Edition, 1987.

(5) I.J. Nagrath, M.Gopal, *Control Systems Engineering*, New Age International Publishers, New Delhi, 4th Edition, 2005.

(6) Katsuhiko Ogata, *Modern Control Engineering*, Prentice-Hall, Upper Saddle River, New Jersey, 3rd Edition, 1997.

(7) M. Gopal, *Modern Control Systems Theory*, New Age International Publishers, New Delhi, 2nd Edition, 1993.

Chapter 2

ANALYSIS OF DISCRETE-TIME SYSTEMS

2.1 INTRODUCTION

The advances made in microprocessors, microcomputers, and digital signal processors have accelerated the growth of digital control systems theory. The discrete-time systems are dynamic systems in which the system-variables are defined only at discrete instants of time.

The terms **sampled-data control systems**, **discrete-time control systems** and **digital control systems** have all been used **interchangeably** in control system literature. Strictly speaking, sampled-data are pulse-amplitude modulated signals and are obtained by some means of sampling an analog signal. Digital signals are generated by means of digital transducers or digital computers, often in **digitally coded form**. The discrete-time systems, in a broad sense, describe all systems having some form of digital or sampled signals.

Discrete-time systems differ from continuous-time systems in that the signals for a discrete-time systems are in sampled-data form. In contrast to the continuous-time system, the operation of discrete-time systems are described by a set of **difference-equations**. The analysis and design of discrete-time systems may be effectively carried out by use of the z-transform, which was evolved from the Laplace transform as a special form.

2.2 SAMPLED-DATA AND DIGITAL CONTROL SYSTEMS

Figure 2.1(a) shows the block-diagram of a sampled-data control system. The continuous error-signal $e(t)$ is sampled at an interval of time T by means of a sampler. The plant output is also a continuous-signal and it is due to the fact that a hold circuit preceeds the plant.

99

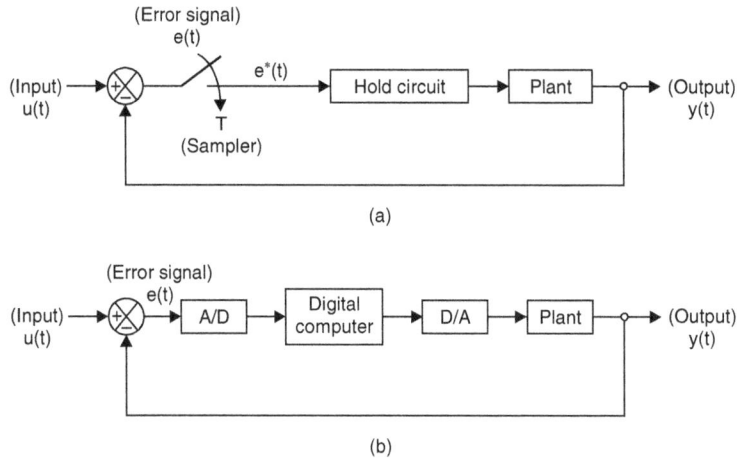

FIGURE 2.1   (a) Block-diagram of a sampled-data control system.
(b) Block-diagram of a digital control system.

Figure 2.1(b) shows the block-diagram of a digital control system. A digital control system uses digital signals and a digital-computer to control a process. The continuous-time error-signals are converted from analog form to digital form by means of an A/D converter. In practice, the A/D converter itself contains the **sampler**. After processing the inputs, the digital computer provides an output in digital form. This output is then converted to analog form by means of a D/A converter. In practice, the D/A converter itself contains the **hold-device**.

The use of sampled data in control systems enables **time-sharing** among different input signals using the same control equipment. Different input signals can be sampled periodically by staggering sampling time and thus a number of inputs can be handled over the same control equipment.

Sampled data technique is most appropriate for control systems requiring long-distance data transmission. It is well known that pulses may be transmitted with little loss of accuracy. Data in analog form may suffer considerable distortion in the transmission channel. Using the sampled-data technique, a number of communication signals can be sequentially sampled and transmitted through a single transmission channel, thus decreasing the cost of transmission installation. This technique is referred to as **time-multiplexing**.

In some control applications, it is required that the load be driven by a weak power input signal. The signal-sampling reduces the power demand made on the signal and is, therefore, helpful for signals of weak power origin.

2.2.1   Sampler

A sampler is the basic element of a discrete-time control system, which converts a continuous-time signal into a train of pulses occuring at the sampling instants 0, T, $2T$, ..., where T is the sampling period. Note that, in between the sampling instants, the sampler does not transmit any information.

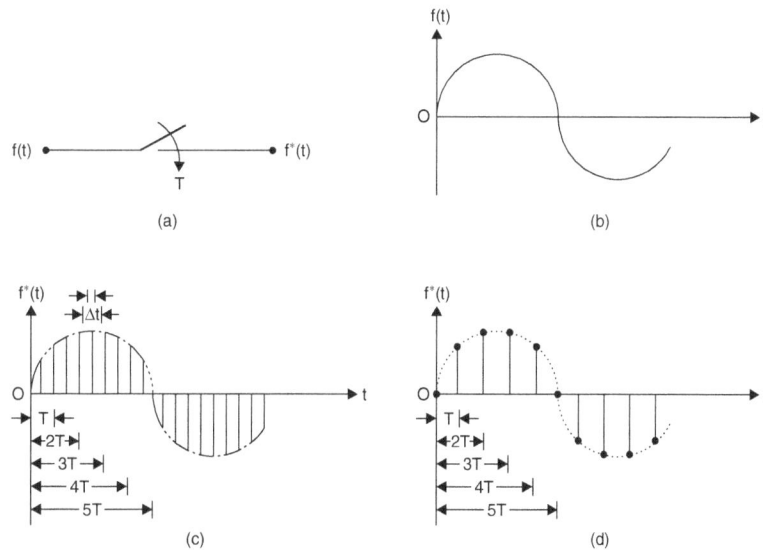

FIGURE 2.2   Uniform periodic sampling:
(a) Sampler   (b) Continuous signal
(c) Practical sampling   (d) Ideal sampling

Figure 2.2(a) shows a switch being used as a **sampler**. A continuous-time signal $f(t)$, as shown in Figure 2.2(b), is the input to the sampler. The switch is closed for a very short duration of time Δt, and then remains open for some duration of time.

The process is repeated with a time period T, called **sampling time**, and the signal at the output end of the sampler is available in the form of pulses of very short duration Δt, each followed by a skip period when no output appears at the output end of the sampler. The continuous-time signal $f(t)$ is thus sampled at a regular interval of time T, as shown in Figure 2.2(c). The sampled signal is denoted as $f^*(t)$, which is obtained after sampling the continuous time input signal $f(t)$.

Figure 2.2(c) shows the **signal-modulated pulse train** having pulse-width Δt and is a case of **practical-sampling**. However, in the case of **ideal-sampling**, the pulse width Δt approaches to zero and therefore the output $f^*(t)$ of an ideal sampler is the **signal-modulated impulse train** as shown in Figure 2.2(d).

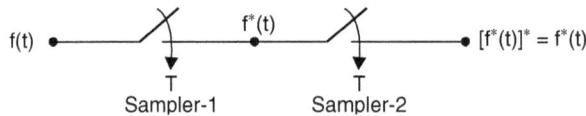

FIGURE 2.3   Signal f(t) sampled twice.

Figure 2.3 shows two samplers **operating in synchronism** i.e., their sampling time and instants of operation are the same. The continuous-time signal $f(t)$ is thus being sampled twice. As the two samplers operate in synchronism, the output of the second sampler will obviously be $f^*(t)$.

Therefore, $$[f^*(t)]^* = f^*(t).$$

Thus, starring a **starred function** results in the starred function itself.

2.2.2 Sampling Process

The sampler converts a continuous-time signal $f(t)$ into a sequence of pulses wherein the magnitude of the pulse gives the strength of the input signal at the instant of sampling. The pulse considered is of very short time duration and thus may be approximated as an **impulse**.

Considering the sampler output to be a train of **weighted impulses**, we can relate the continuous-time signal $f(t)$ to the sampler output $f^*(t)$ as

$$f^*(t) = f(t)\delta_T(t) \tag{2.1}$$

where $\delta_T(t)$ represents a train of **unit impulses**.

The sampler output is equal to the product of the continuous-time input signal $f(t)$ and the train of unit-impulses $\delta_T(t)$. In other words, the sampler may be considered as a **modulator** with the continuous-time input signal $f(t)$ as the **modulating signal** and the train of unit-impulses $\delta_T(t)$ as the **carrier**, as shown in Figure 2.4.

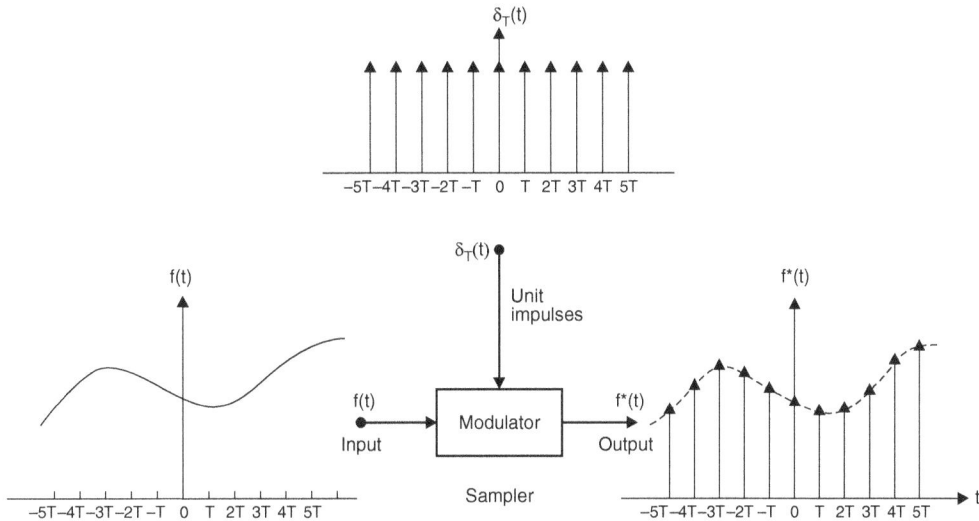

FIGURE 2.4   Sampling process.

The unit-impulse train may be expressed mathematically as

$$\delta_T(t) = \sum_{k=-\infty}^{\infty} \delta(t-kT) \tag{2.2}$$

where $\delta(t - kT)$ is the unit-impulse occuring at $t = kT$.

As the continuous-time signal $f(t)$ appears only at sampling instants in the output; the same is denoted by $f(kT)$ and the sampled-output signal can be represented as

$$f^*(t) = \sum_{k=-\infty}^{\infty} f(kT)\delta(t-kT) \tag{2.3}$$

Equation (2.3) gives the **sampler output**, which is a train of weighted impulses.

Most of the standard time-functions are zero for $t < 0$, then in Equation (2.3), the lower limit can be put as zero instead of $-\infty$ as the function $f(kT)$ is zero for $t < 0$. In this book, therefore, unless otherwise stated, we shall assume that this is the case.

Thus, Equation (2.3) becomes

$$f^*(t) = \sum_{k=0}^{\infty} f(kT)\ \delta(t - kT).$$

(2.4)

2.2.2.1 Laplace Transform of a Sampled (Starred) Function

The unit-impulse $\delta(t - kT)$, occuring at sampling instants $t = kT$; $k = 0$, 1, 2, ..., is the first time-derivative of the sampled unit-step function $u(t - kT)$ occuring at $t = kT; k = 0, 1, 2, ...$.

Therefore,

$$£\ \delta(t - kT) = £\frac{d}{dt}[u(t - kT)]$$

$$= s \cdot \frac{1}{s} e^{-skT}$$

or $\qquad £\ \delta(t - kT) = e^{-skT}.$

Thus, the Laplace transform of Equation (2.4) is given by

$$F^*(s) = \sum_{k=0}^{\infty} f(kT)\ e^{-skT}.$$

(2.5)

$F^*(s)$ is known as the Laplace transform of a sampled (starred) function or simply a **starred transform**.

If $f(t)$ is a unit-step function, then $f(kT) = 1$ and the Laplace transform of the starred unit-step function $f^*(t)$ is given by

$$F^*(1) = \sum_{k=0}^{\infty} 1 \cdot e^{-skT}$$

or, $\qquad F^*(1) = 1 + e^{-sT} + e^{-2sT} + ...$

or, $\qquad F^*(1) = \dfrac{1}{1 - e^{-sT}}.$

(2.6)

2.2.3   Signal Reconstruction: Holding Devices

The use of a sampler in control systems generates high-frequency components in the sampled ouput. The sampler input **Fourier-spectra** is shown in Figure 2.5(b) wherein ω_m is the highest frequency contained in the input

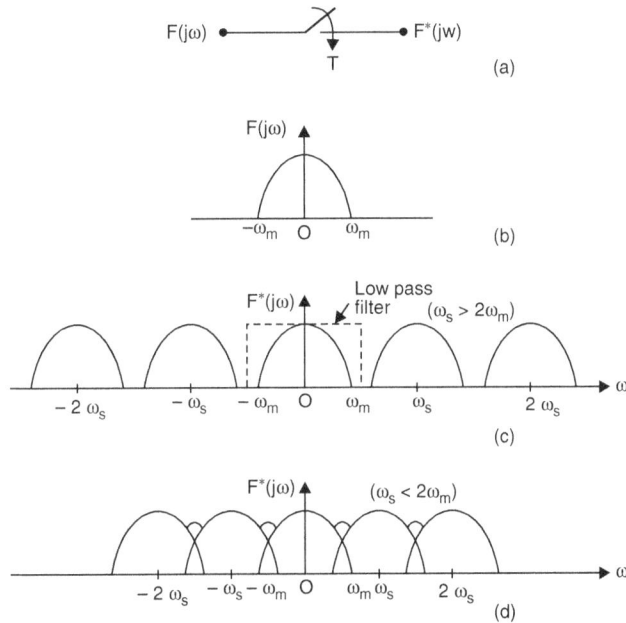

FIGURE 2.5

(*a*) Sampler   (*b*) Sampler input Fourier spectra
(*c*) Sampler output Fourier   (*d*) Sampler output Fourier
    spectra: $\omega_s > 2\omega_m$*   spectra: $\omega_s < 2\omega_m$*

signal. The sampler output contains the original input frequency component ω_m and complementary high-frequency components in addition, as shown in Figure 2.5(*c*) and Figure 2.5(*d*) that correspond to $\omega_s > 2\omega_m$ and $\omega_s < 2\omega_m$ respectively; where ω_s is the **sampling frequency**.

It is immediately observed from Figure 2.5(*c*) and Figure 2.5(*d*) that so long as $\omega_s \geq 2\,\omega_m$, the original spectrum is preserved in the sampled signal and can be extracted from it by **low-pass filtering** (shown in the dotted line in Figure 2.5 (*c*)). This is the well-known **Shanon's sampling theorem** according to which the information contained in a signal is fully preserved in the sampled version so long as the sampling frequency (ω_s) is at least twice the maximum frequency (ω_m) contained in the signal. That is,

$$\omega_s \geq 2\omega_m.$$

The original signal is reconstructed from the sampled signal by means of various types of **hold-circuits** (**extrapolators**). The simplest hold-circuit is the **zero-order-hold** (**ZOH**) in which the reconstructed signal acquires the same value as the last received sample for the entire sampling period. The zero-order-hold (ZOH) circuit has the characteristics of a low-pass filter and therefore, if a ZOH circuit is placed after the sampler, the high-frequency (complementary) components present in the reconstructed signal are easily filtered out and at the output of ZOH the original signal appears. The ZOH device is also known as a **"box-car" generator**. The schematic diagram of sampler and ZOH is shown in Figure 2.6(*a*), while signal reconstruction is illustrated in Figures 2.6(*b*) and 2.6(*c*).

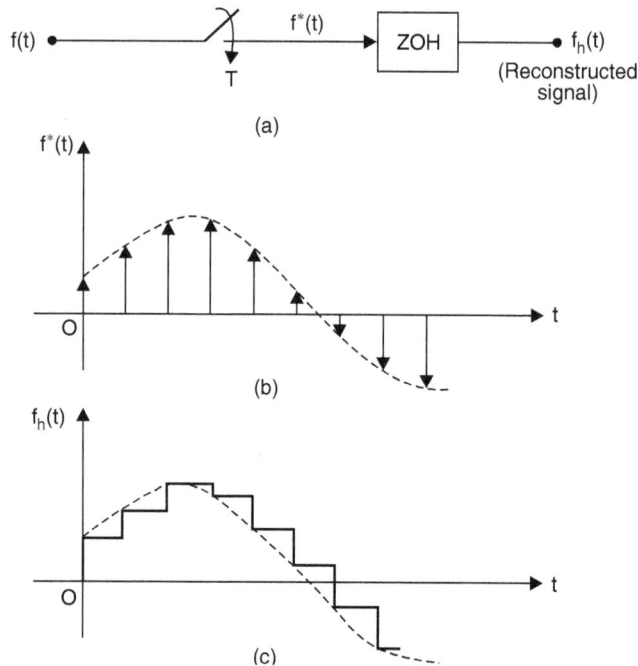

FIGURE 2.6 (*a*) Sampler and zero-order-hold (ZOH). (*b*) Sampled signal. (*c*) Reconstructed signal by ZOH.

Note that the higher-order holds offer no particular advantage over the ZOH. The ZOH, when used in conjunction with a high sampling rate, provides satisfactory performance.

2.2.3.1 Transfer-Function of a ZOH Circuit

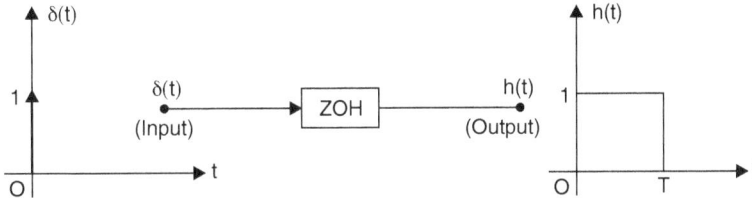

FIGURE 2.7   **Unit impulse input to ZOH.**

In Figure 2.7, a unit impulse input is given to a ZOH circuit that holds the input signal for a duration T and thus the output appears to be a unit-step function until duration T. Therefore, the output of ZOH can be given as

$$h(t) = u(t) - u(t - T).$$

Therefore, the Laplace transform of the output of ZOH is given as

$$H(s) = \frac{1}{s} - \frac{1}{s} . e^{-sT}$$

or,
$$H(s) = \frac{(1 - e^{-sT})}{s}.$$

Now, the transfer function of ZOH is given by

$$G_{ho}(s) = \frac{\text{Laplace transform of output of ZOH}}{\text{Laplace transform of input to ZOH}}$$

or,
$$G_{ho}(s) = \frac{H(s)}{1}$$

or,
$$G_{ho}(s) = \frac{(1 - e^{-sT})}{s}. \tag{2.7}$$

2.2.3.2 Frequency Response of ZOH

We already know from Equation (2.7) that the transfer function of a zero-order-hold circuit is

$$G_{ho}(s) = \frac{(1 - e^{-sT})}{s}.$$

Put $s = j\omega$ in the preceeding relation and we have

$$G_{ho}(j\omega) = \frac{(1 - e^{-j\omega T})}{j\omega}$$

$$= \frac{(1 - e^{-j\omega T})}{j\omega} . e^{j\omega T/2} . e^{-j\omega T/2}$$

$$= \frac{(e^{j\omega T/2} - e^{-j\omega T/2})}{j\omega} . e^{-j\omega T/2}$$

$$= \left[\frac{e^{j\omega T/2} - e^{-j\omega T/2}}{2j} \right] . \frac{2e^{-j\omega T/2}}{\omega}$$

$$= \sin\left(\frac{\omega T}{2}\right) . \frac{2e^{-j\omega T/2}}{\omega}$$

$$= \frac{Te^{-j\omega T/2}}{\left(\dfrac{\omega T}{2}\right)} . \sin\left(\frac{\omega T}{2}\right)$$

or,
$$G_{ho}(j\omega) = \frac{T \sin\left(\dfrac{\omega T}{2}\right)}{\left(\dfrac{\omega T}{2}\right)} . e^{-j\omega T/2}.$$

If the sampling frequency is ω_s, then the sampling time T is given by

$$T = \frac{2\pi}{\omega_s}$$

then,
$$G_{ho}(j\omega) = \frac{T \sin\left(\dfrac{\pi\omega}{\omega_s}\right)}{\left(\dfrac{\pi\omega}{\omega_s}\right)} \cdot e^{-j\left(\dfrac{\pi\omega}{\omega_s}\right)}.$$

Thus,
$$|G_{ho}(j\omega)| = T \left| \frac{\sin\left(\dfrac{\pi\omega}{\omega_s}\right)}{\left(\dfrac{\pi\omega}{\omega_s}\right)} \right| \qquad (2.8\ A)$$

and,
$$\angle G_{ho}(j\omega) = -\left(\frac{\pi\omega}{\omega_s}\right) + \theta$$

where,
$$\theta = \begin{cases} 0, & \sin\left(\dfrac{\pi\omega}{\omega_s}\right) > 0 \\[4mm] \pi, & \sin\left(\dfrac{\pi\omega}{\omega_s}\right) < 0 \end{cases} \qquad (2.8\ B)$$

To view the previous derivations, the amplitude and phase plots are shown in Figure 2.8.

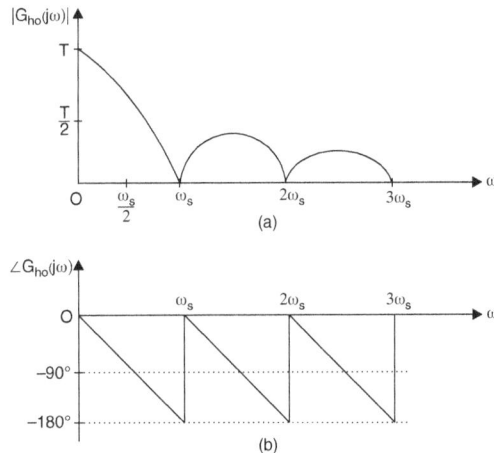

FIGURE 2.8 **Frequency response of the ZOH: (*a*) Amplitude plot (*b*) Phase plot.**

2.3   THE z-TRANSFORM

The z-transform is a powerful mathematical tool for analysis of discrete-time systems just as the Laplace-transform is for continuous-time systems.

The **two-sided z-transform** of a discrete-time signal $f(k)$ is defined as

$$F(z) = Z[f(k)] = \sum_{k=-\infty}^{\infty} f(k)z^{-k} \qquad (2.9\ A)$$

where z is a **complex variable** with real and imaginary parts. The two-sided z-transform is also called a **bilateral z-transform**.

The **one-sided z-transform** of a discrete-time signal $f(k)$ is defined as

$$F(z) = Z[f(k)] = \sum_{k=0}^{\infty} f(k)z^{-k}. \qquad (2.9\ B)$$

It is also called a **unilateral z-transform**.

The difference between a two-sided z-transform and a one-sided z-transform is that the lower limit of summation is zero for a one-sided z-transform and $-\infty$ for a two-sided z-transform.

For **causal systems** (a system whose present output depends only on present and past inputs, not on future values of the input); the two-sided and one-sided z-transforms are equivalent. One-sided z-transforms are used for solving linear difference equations with nonzero initial conditions.

The portion of the z-plane for which the series in Equations (2.9) converges is called the **region of convergence(ROC)**. For some values of z, the power series in Equations (2.9) may not converge to a finite value. Thus, the ROC depends upon the value of z. The possible configurations of the ROC for a z-transform may be:

1. The entire z-plane [Figure 2.9(a)]
2. Interior of a circle [Figure 2.9(b)]
3. Exterior of a circle [Figure 2.9(c)]
4. An annulus [Figure 2.9(d)]

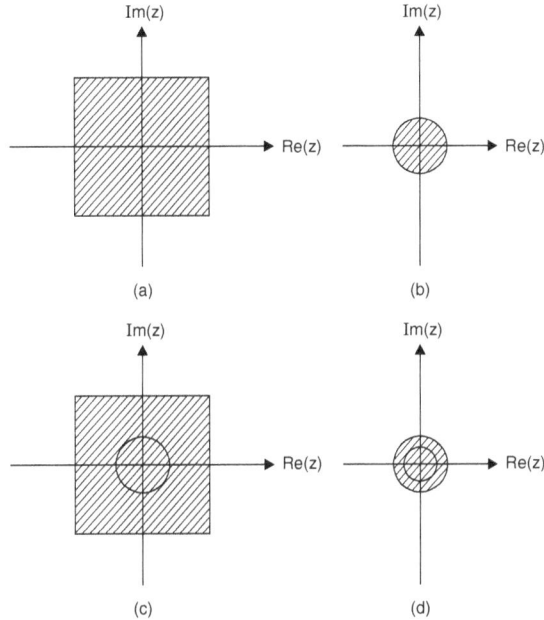

FIGURE 2.9 Possible configurations of the ROC for the z-transform:
(*a*) The entire z-plane (*b*) Interior of a circle
(*c*) Exterior of a circle (*d*) An annulus

For most of the physical systems, $f(k) = 0; k < 0$, therefore, unless otherwise stated, we shall consider only the case of the **one-sided z-transform**, defined by Equation (2.9*b*).

2.3.1 Properties of z-transform

In this section, we will study some of the important properties of the z-transform.

2.3.1.1 Linearity

$$\mathcal{Z}[af(k) + bg(k)] = \sum_{k=0}^{\infty} [af(k) + bg(k)]z^{-k}$$

$$= a\sum_{k=0}^{\infty} f(k)z^{-k} + b\sum_{k=0}^{\infty} g(k)z^{-k}$$

$$= aF(z) + bG(z) \qquad (2.10)$$

2.3.1.2 Time-shifting

Consider the sequence,

$$g(k) = f(k+1); \ k \geq -1,$$

which is the sequence $f(k)$ shifted one interval to the left (advanced). Taking the z-transform,

$$\mathcal{Z}[f(k+1)] = \sum_{k=0}^{\infty} f(k+1)z^{-k}$$

$$= z\sum_{k=0}^{\infty} f(k+1)z^{-(k+1)}.$$

Let $k + 1 = m$, and we have

$$\mathcal{Z}[f(k+1)] = z\sum_{m=1}^{\infty} f(m)\, z^{-m}$$

$$= z\left[\sum_{m=0}^{\infty} f(m)\, z^{-m} - f(0)\right]$$

$$= zF(z) - zf(0). \tag{2.11}$$

In general,

$$\mathcal{Z}[f(k+n)] = z^n F(z) - \sum_{i=0}^{n-1} f(i)\, z^{n-i}; k \geq -n. \tag{2.12}$$

Now, consider the sequence,

$$g(k) = f(k-1); \ k \geq 1,$$

which is the sequence $f(k)$ shifted one interval to the right (delayed). Taking the z-transform

$$\mathcal{Z}[f(k-1)] = \sum_{k=0}^{\infty} f(k-1)\, z^{-k}$$

$$= z^{-1}\sum_{k=0}^{\infty} f(k-1)\, z^{-(k-1)}.$$

Let $k - 1 = m$, and we have

$$\mathcal{Z}[f(k-1)] = z^{-1}\sum_{m=-1}^{\infty} f(m)\, z^{-m}.$$

As $f(m) = 0$ for $m < 0$, we have

$$\mathcal{Z}[f(k-1)] = z^{-1}\sum_{m=0}^{\infty} f(m)\, z^{-m}$$

$$= z^{-1}F(z). \tag{2.13}$$

In general,

$$\mathcal{Z}[f(k-n)] = z^{-n}F(z); k \geq n. \tag{2.14}$$

2.3.1.3 Scaling in z-domain

Let
$$\mathcal{Z}[f(k)] = F(z).$$

Then,
$$\mathcal{Z}[a^{k}f(k)] = \sum_{k=0}^{\infty} a^{k}f(k)\, z^{-k}$$

$$= \sum_{k=0}^{\infty}\left(\frac{1}{a}\right)^{-k} f(k)\, z^{-k}$$

$$= \sum_{k=0}^{\infty} f(k)\left(\frac{z}{a}\right)^{-k}$$

$$= F\left(\frac{z}{a}\right). \tag{2.15}$$

2.3.1.4 Differentiation in z-domain

Let
$$\mathcal{Z}[f(k)] = F(z).$$

Then,
$$\mathcal{Z}[kf(k)] = \sum_{k=0}^{\infty} kf(k)\, z^{-k}$$

$$= -z\sum_{k=0}^{\infty}[-kf(k)]\, z^{-k-1}$$

$$= -z \sum_{k=0}^{\infty} f(k) \frac{d}{dz} z^{-k}$$

$$= -z \frac{d}{dz} \sum_{k=0}^{\infty} f(k) \, z^{-k}$$

$$= -z \frac{d}{dz} F(z). \qquad (2.16)$$

In general,

$$\mathcal{Z}[k^n f(k)] = \left(-z \frac{d}{dz}\right)^n F(z). \qquad (2.17)$$

2.3.1.5 Convolution

The convolution property is one of the most important properties of the z-transform because it is used to convert the convolution of two discrete-time signals in time-domain into a multiplication of their z-transforms.

Let $\mathcal{Z}[f(k)] = F(z)$

and, $\mathcal{Z}[g(k)] = G(z)$.

Then, $\mathcal{Z}[f(k) * g(k)] = F(z) \, G(z)$.   (2.18)

In order to prove the preceeding property, students may refer section 2.6.

2.3.1.6 Initial Value Theorem

$$\mathcal{Z}[f(k)] = F(z) = \sum_{k=0}^{\infty} f(k) z^{-k}$$

or, $F(z) = f(0) + f(1) \, z^{-1} + f(2) \, z^{-2} + \dots.$

Taking the limit $z \to \infty$ as $k \to 0$, we get

$$f(0) = \lim_{z \to \infty} F(z). \qquad (2.19)$$

2.3.1.7 Final Value Theorem

$$\mathcal{Z}f(k) = F(z) = \sum_{k=0}^{\infty} f(k)z^{-k}$$

Now, $\mathcal{Z}[f(k+1)] = \sum_{k=0}^{\infty} f(k+1)\,z^{-k}.$ (2.20 A)

Also, $\mathcal{Z}[f(k+1)] = zF(z) - zf(0)$ (Property 2.11).   (2.20 B)

From Equations (2.20) we may write

$$zF(z) - zf(0) = \sum_{k=0}^{\infty} f(k+1)z^{-k}$$

or, $zF(z) - zf(0) - F(z) = \sum_{k=0}^{\infty} f(k+1)z^{-k} - \sum_{k=0}^{\infty} f(k)z^{-k}$

or, $(z-1)\,F(z) - zf(0) = \sum_{k=0}^{\infty} [f(k+1) - f(k)]z^{-k}$

or, $(z-1)\,F(z) = zf(0) + \sum_{k=0}^{\infty} [f(k+1) - f(k)]\,z^{-k}.$

For the final value theorem to be applicable, $F(z)$ must not have any pole in the region outside a unit circle i.e., it should be **analytic** for $|z| > 1$.

Because of the assumed stability condition stated previously, taking the limit $z \to 1$ as $k \to \infty$, we obtain

$$\operatorname*{Lim}_{z \to 1}[(z-1)\,F(z)] = f(0) + f(\infty) - f(0)$$

or, $f(\infty) = \operatorname*{Lim}_{z \to 1}[(z-1)\,F(z)].$ (2.21)

Provided $F(z)$ is **analytic** for $|z| > 1$.

2.3.2   z-Transforms of Some Common Discrete-time-sequences

2.3.2.1 Unit-Delta Function

It is defined as

$$\delta(k) = 1 \quad \text{for } k = 0$$
$$= 0 \quad \text{for } k \neq 0.$$

Using definition (2.9b), we have

$$\mathcal{Z}[\delta(k)] = \sum_{k=0}^{\infty} \delta(k)\, z^{-k}$$
$$= \delta(0)\, z^0 + \delta(1)z^{-1} + \delta(2)\, z^{-2} + ...$$
$$= 1. \qquad\qquad\qquad (2.22)$$

2.3.2.2 Discrete Unit-Step Function

It is defined as

$$u(k) = 1 \qquad \text{for } k \geq 0$$
$$= 0 \qquad \text{for } k < 0.$$

Using definition (2.9b), we have

$$\mathcal{Z}[u(k)] = \sum_{k=0}^{\infty} u(k)\, z^{-k}$$
$$= 1.z^0 + 1.z^{-1} + 1.z^{-2} + ...$$
$$= 1 + z^{-1} + z^{-2} + ...$$
$$= \frac{1}{1 - z^{-1}}$$
$$= \frac{z}{z-1} \qquad\qquad \text{ROC: } |z| > 1. \qquad (2.23)$$

2.3.2.3 Discrete Unit Ramp Function

A discrete unit ramp is obtained by multiplying a discrete unit step function by k.

i.e., $\qquad\qquad r(k) = ku\,(k)$

Then, $\qquad \mathcal{Z}[r(k)] = \mathcal{Z}[ku(k)]$

$$= -z \frac{d}{dz} U(z). \qquad\qquad \text{(Property 2.16)}$$

Since $\qquad U(z) = \dfrac{z}{z-1}.$

Therefore, $\quad \mathcal{Z}[r(k)] = -z\dfrac{d}{dz}\left(\dfrac{z}{z-1}\right)$

$$= \dfrac{z}{(z-1)^2} \qquad \text{ROC: } |z| > 1. \qquad (2.24)$$

2.3.2.4 Discrete Sine and Cosine Functions

Consider an exponential signal

$$f(kT) = e^{j\omega kT} = \cos \omega kT + j \sin \omega kT$$

where, T is the sampling interval.

Let $\qquad e^{j\omega T} = a$

then, $\qquad f(kT) = a^k.$

Thus, $\qquad \mathcal{Z}[f(kT)] = \mathcal{Z}[a^k]$

$$= \sum_{k=0}^{\infty} a^k z^{-k}$$

$$= 1 + \dfrac{a}{z} + \dfrac{a^2}{z^2} + \dots\dots$$

$$= \dfrac{1}{1 - \dfrac{a}{z}} = \dfrac{z}{z - a}$$

$$= \dfrac{z}{z - e^{j\omega T}}$$

$$= \dfrac{z}{z - (\cos \omega T + j \sin \omega T)}$$

$$= \dfrac{z}{(z - \cos \omega T) - j \sin \omega T}$$

$$= \dfrac{z[(z - \cos \omega T) + j \sin \omega T]}{(z - \cos \omega T)^2 + \sin^2 \omega T}$$

$$= \dfrac{z^2 - z \cos \omega T + jz \sin \omega T}{z^2 + \cos^2 \omega T - 2z \cos \omega T + \sin^2 \omega T}$$

$$= \frac{z^2 - z \cos \omega T + jz \sin \omega T}{z^2 - 2z \cos \omega T + (\cos^2 \omega T + \sin^2 \omega T)}$$

$$= \frac{z^2 - z \cos \omega T}{z^2 - 2z \cos \omega T + 1} + j \frac{z \sin \omega T}{z^2 - 2z \cos \omega T + 1}.$$

Expanding the L.H.S. of the previous equation, we get

$$\mathcal{Z}[\cos \omega kT + j \sin \omega kT] = \frac{z^2 - z \cos \omega T}{z^2 - 2z \cos \omega T + 1} + j \frac{z \sin \omega T}{z^2 - 2z \cos \omega T + 1}.$$

Equating the real and imaginary parts, we get

$$\mathcal{Z}[\cos \omega kT] = \frac{z^2 - z \cos \omega T}{z^2 - 2z \cos \omega T + 1}$$

$$\mathcal{Z}[\sin \omega kT] = \frac{z \sin \omega T}{z^2 - 2z \cos \omega T + 1}.$$

Example 2.1. *Determine the z-transform of the following discrete-time signals*

$(a) f(k) = u(k-1)$ $\qquad\qquad$ $(b) f(k) = -a^k u(-k-1).$

Solution.

(a) Given, $\qquad\qquad\qquad\qquad f(k) = u(k - 1).$

Using Property (2.13), we have

$$\mathcal{Z}[f(k)] = \mathcal{Z}[u(k-1)] = z^{-1}U(z).$$

Since, $\qquad U(z) = \dfrac{z}{z-1}.$

Therefore,

$$\mathcal{Z}[f(k)] = \mathcal{Z}[u(k-1)]$$

$$= z^{-1}\left(\frac{z}{z-1}\right)$$

$$= \frac{1}{z-1} \qquad\qquad \text{ROC: } |z| > 1 \ \textbf{Ans.}$$

(b) Given, $\quad f(k) = -a^k u(-k-1)$.

The discrete unit-step function is defined as

$$u(k) = 1 \quad \text{for } k \geq 0$$
$$= 0 \quad \text{for } k < 0.$$

Thus, $\quad u(-k-1) = 1 \quad \text{for } k \leq -1$
$$= 0 \quad \text{for } k > 0.$$

Therefore, $\quad \mathcal{Z}[f(k)] = \mathcal{Z}[-a^k u(-k-1)]$

$$= \sum_{k=-\infty}^{\infty} [-a^k \, u\,(-k-1)\, z^{-k}$$

$$= \sum_{k=-\infty}^{-1} (-a^k)\, z^{-k}$$

$$= -\sum_{k=-\infty}^{-1} (a^{-1}z)^{-k}.$$

Let $\quad m = -k.$

Then, $\quad \mathcal{Z}[f(k)] = -\sum_{m=\infty}^{1} (a^{-1}z)^m$

$$= \sum_{m=1}^{\infty} (a^{-1}z)^m$$

$$= \left(\frac{z}{a}\right) + \left(\frac{z}{a}\right)^2 + \left(\frac{z}{a}\right)^3 + \dots..$$

$$= \frac{\left(\dfrac{z}{a}\right)}{1 - \left(\dfrac{z}{a}\right)}$$

$$= \frac{z}{a-z} \qquad\qquad \text{ROC: } |z| < |a|.$$

The ROC is **interior** of a circle having radius $|a|$. **Ans.**

Example 2.2. *Determine the z-transform of the functions defined as*

$(a) f(k) = \left(\dfrac{1}{a}\right)^k$ *for* $k \geq 0$

$\qquad = 0 \quad$ *for* $k < 0$

$(b) f(k) = \dfrac{a^k}{k!}.$

Solution. (a) Given, $\quad f(k) = \left(\dfrac{1}{a}\right)^k$ for $k \geq 0$

$\qquad\qquad\qquad = 0 \quad$ for $k < 0$

$\therefore \qquad\qquad \mathcal{Z}[f(k)] = \displaystyle\sum_{k=0}^{\infty} \left(\dfrac{1}{a}\right)^k z^{-k}$

$$= 1 + \dfrac{1}{az} + \dfrac{1}{a^2 z^2} + \ldots\ldots$$

$$= \dfrac{1}{1 - \dfrac{1}{az}}$$

$$= \dfrac{az}{az - 1} \qquad\qquad \text{ROC: } |z| > \left|\dfrac{1}{a}\right| \textbf{ Ans.}$$

(b) Given,

$$f(k) = \dfrac{a^k}{k!}$$

$\therefore \qquad\qquad \mathcal{Z}[f(k)] = \displaystyle\sum_{k=0}^{\infty} \left(\dfrac{a^k}{k!}\right) z^{-k}$

$$= \displaystyle\sum_{k=0}^{\infty} \dfrac{1}{k!} \left(\dfrac{a}{z}\right)^k$$

$$= 1 + \left(\dfrac{a}{z}\right) + \dfrac{1}{2!}\left(\dfrac{a}{z}\right)^2 + \ldots\ldots$$

$$= \exp\left(\dfrac{a}{z}\right) \qquad\qquad \text{ROC} : |z| > 0.$$

The ROC is the **entire z-plane** except the origin. **Ans.**

2.4   THE INVERSE z-TRANSFORM

The inverse z-transformation is a process of determining the sequence that generates a given z-transform. Thus, the mechanism for transforming $F(z)$ back to a discrete-time sequence $f(k)$ is called the inverse z-transform.

Three methods are often used for the evaluation of the inverse z-transform in practice. These methods are

1. Power-Series Expansion Method
2. Partial-Fraction Expansion Method
3. Contour-Integration Method

We shall study the first two methods. The third method of contour-integration is beyond the scope of this book.

2.4.1   Power-Series Expansion Method

The z-transform $F(z)$ of a discrete-time sequence $f(k)$ can mostly be expressed in the ratio of polynomial form as

$$F(z) = \frac{b_0 z^m + b_1 z^{m-1} + \ldots\ldots + b_m}{z^n + a_1 z^{n-1} + \ldots\ldots + a_n} ; m \le n \ \text{(For causal systems)}. \qquad (2.27)$$

We can immediately recognise the discrete-time sequence $f(k)$ if $F(z)$ can be written in a form of series with increasing powers of z^{-1}. This is easily carried out by dividing the numerator by the denominator **(by long division process)**.

Example 2.3. *Find the sequence f(k) when F(z) is given by*

$$F(z) = \frac{10z}{(z-1)(z-2)}$$

Solution. $F(z) = \dfrac{10z}{(z-1)(z-2)}$ or $\qquad F(z) = \dfrac{10z}{z^2 - 3z + 2}$

Long division is performed as

$$
z^2 - 3z + 2 \overline{\smash{\big)}\ 10z \phantom{+ 30z^{-2} + 70z^{-3}}} \quad 10z^{-1} + 30z^{-2} + 70z^{-3} + \ldots\ldots
$$

$$10z - 30 + 20\,z^{-1}$$

$$-\quad +\quad -$$

$$30 - 20\,z^{-1}$$

$$30 - 90z^{-1} + 60z^{-2}$$

$$-\quad +\quad -$$

$$70z^{-1} - 60z^{-2}$$

$$70z^{-1} - 210z^{-2} + 140\,z^{-3}$$

$$-\quad +\quad -$$

$$150z^{-2} - 140z^{-3}$$

$$.$$
$$.$$
$$.$$

and so on.

Thus, $F(z) = 10z^{-1} + 30z^{-2} + 70z^{-3} + \ldots\ldots$.

But $F(z) = f(0)z^0 + f(1)z^{-1} + f(2)z^{-2} + f(3)z^{-3} + \ldots\ldots$.

On comparison, we have

$$f(0) = 0.$$
$$f(1) = 10.$$
$$f(2) = 30.$$
$$f(3) = 70.$$

Therefore,

$$f(k) = \{0, 10, 30, 70, \ldots\ldots\}.$$ **Ans.**

From the previous example it is observed that the power-series expansion method gives $f(k)$ for any k, but it does not help us to get the general form of $f(k)$. Therefore, it is useful only in numerical studies and not for analytical ones.

2.4.2   Partial-Fraction Expansion Method

Let us assume that the z-transform $F(z)$ of a discrete-time sequence $f(k)$ is given as

$$F(z) = A_0 + \frac{M(z)}{N(z)}$$

where the polynomial $M(z)$ is of an order less than that of $N(z)$. Assuming **distinct poles** of $F(z)$, we can expand $F(z)$ into partial-fractions as

$$F(z) = A_0 + \frac{A_1}{z - a_1} + \frac{A_2}{z - a_2} + \ldots\ldots + \frac{A_n}{z - a_n}. \tag{2.28}$$

Now, $\mathcal{Z}^{-1}\left[\dfrac{1}{z - a_i}\right] = \mathcal{Z}^{-1}\left[z^{-1}\left(\dfrac{z}{z - a_i}\right)\right]$

$$= \mathcal{Z}^{-1}\left[z^{-1}\left\{\mathcal{Z}(a_i)^k\right\}\right]$$

$$= \mathcal{Z}^{-1}\left[\mathcal{Z}(a_i)^{k-1}\right]$$

$$= (a_i)^{k-1}; k \geq 1. \tag{Property 2.13}$$

Hence, taking the inverse z-transform of Equation (2.28), we get

$$f(k) = A_0 \delta(k) + A_1 a_1^{k-1} + A_2 a_2^{k-1} + \ldots\ldots + A_n a_n^{k-1}; k \geq 1. \tag{2.29}$$

In case of **repeated roots**, we get the factors of the type $\dfrac{1}{(z - a)^2}$ in the partial fraction expansion. For this case

$$\mathcal{Z}^{-1}\left[\frac{1}{(z - a)^2}\right] = \mathcal{Z}^{-1}\left[\frac{1}{a} z^{-1}\left\{\frac{az}{(z - a)^2}\right\}\right]$$

$$= \mathcal{Z}^{-1}\left[\frac{1}{a} z^{-1}\left\{\mathcal{Z}(ka^k)\right\}\right]$$

$$= \mathcal{Z}^{-1}\left[\frac{1}{a} \mathcal{Z}\left\{(k-1)a^{k-1}\right\}\right] \tag{Property 2.13}$$

$$= \mathcal{Z}^{-1}[\mathcal{Z}\{(k-1)\,a^{k-2}\}]$$

$$= (k-1)\,a^{k-2}; k \geq 1.$$

This method is illustrated in Example 2.4 and Example 2.5.

Example 2.4. *Find f(k) when F(z) is given by*

$$F(z) = \frac{10z}{(z-1)(z-2)}.$$

Solution. $F(z) = \dfrac{10z}{(z-1)(z-2)}$

$$= \frac{10z}{(z-2)}\bigg|_{z=1} \frac{1}{(z-1)} + \frac{10z}{(z-1)}\bigg|_{z=2} \frac{1}{(z-2)}$$

$$= \frac{-10}{(z-1)} + \frac{20}{(z-2)}$$

Taking the inverse z-transform, we have

$$\begin{aligned} f(k) &= -10(1)^{k-1} + 20(2)^{k-1}; k \geq 1 \\ &= -10 + 10(2)^k \\ &= 10(-1 + 2^k) \\ &= 10(2^k - 1) \end{aligned}$$

Ans.

In order to check the result with $f(k)$ obtained in Example (2.3), putting $k = 0, 1, 2, 3, \ldots\ldots$ we have

$$\begin{aligned} f(0) &= 0, \\ f(1) &= 10, \\ f(2) &= 30, \\ f(3) &= 70. \end{aligned}$$

which are the same as in Example (2.3).

Example 2.5. *Obtain the inverse z-transform using the partial-fraction expansion method given that*

$$F(z) = \frac{4z^2 - 2z}{z^3 - 5z^2 + 8z - 4}.$$

Solution. $F(z) = \dfrac{4z^2 - 2z}{z^3 - 5z^2 + 8z - 4}$

$$= \dfrac{4z^2 - 2z}{(z-1)(z-2)^2}$$

$$= \dfrac{4z^2 - 2z}{(z-2)^2}\bigg|_{z=1} \dfrac{1}{(z-1)} + \dfrac{4z^2 - 2z}{(z-1)}\bigg|_{z=2} \dfrac{1}{(z-2)^2} + \dfrac{d}{dz}\left[\dfrac{4z^2 - 2z}{(z-1)}\right]\bigg|_{z=2} \dfrac{1}{(z-2)}$$

$$= \dfrac{2}{(z-1)} + \dfrac{12}{(z-2)^2} + \dfrac{2}{(z-2)}$$

Taking the inverse z-transform, we get

$$f(k) = 2(1)^{k-1} + 12\,(k-1)(2)^{k-2} + 2(2)^{k-1};\ k \ge 1$$
$$= 2(1)^{k-1} + 6\,(k-1)(2)^{k-1} + 2(2)^{k-1} \qquad\qquad \textbf{Ans.}$$

Example 2.6. *Determine the discrete-time sequence for which the z-transform is given by*

$$F(z) = \log(1 + az^{-1});\ \textbf{\textit{ROC}} : |z| > |a|.$$

Solution. Given, $F(z) = \log(1 + az^{-1})$

$\therefore$ $$\dfrac{d}{dz} F(z) = \dfrac{-az^{-2}}{1 + az^{-1}}$$

or, $$-z\dfrac{d}{dz} F(z) = \dfrac{az^{-1}}{1 + az^{-1}} = \dfrac{a}{z + a}. \qquad\qquad (2.30)$$

Using Property (2.16), we have

$$-z\dfrac{d}{dz} F(z) = \mathcal{Z}[kf(k)]. \qquad\qquad (2.31)$$

Comparing Equations (2.30) and (2.31), we get

$$\mathcal{Z}[kf(k)] = \dfrac{a}{z + a}.$$

Taking the inverse z-transform

$$kf(k) = \mathcal{Z}^{-1}\left[\frac{a}{z+a}\right]$$

$$= a\mathcal{Z}^{-1}\left[\frac{1}{z+a}\right]$$

$$= a(-a)^{k-1}$$

$$= (-1)^{k-1}a^k$$

or, $$f(k) = (-1)^{k-1}\frac{a^k}{k};\ k \geq 1$$ **Ans.**

2.5   SOLUTION OF DIFFERENCE-EQUATIONS USING THE Z-TRANSFORM

In contrast to the continuous-time system, the operation of discrete-time systems are described by a set of difference-equations. The analysis and design of discrete-time systems may be effectively carried out in terms of difference-equations. Essentially, by using the z-transform method, we can transform difference-equations into algebraic equations in z. In order to solve difference-equation we shall utilize the time-shifting property (2.12) of the z-transform, given as

$$\mathcal{Z}[f(k+n)] = z^n F(z) - \sum_{i=0}^{n-1} f(i)\, z^{n-i};\ k \geq -n.$$

Using the previous equation, we can write

$$\mathcal{Z}[f(k+1)] = zF(z) - zf(0)$$

$$\mathcal{Z}[f(k+2)] = z^2 F(z) - z^2 f(0) - zf(1)$$

$$\mathcal{Z}[f(k+3)] = z^3 F(z) - z^3 f(0) - z^2 f(1) - zf(2)$$

and so on.

It is to be noted that when the difference-equation is transformed into an algebraic equation in z by the z-transform method, the initial data are automatically included in the algebraic representation.

Example 2.7. *Solve the following difference-equation by using the z-transform method*

$$f(k+2)+3f(k+1)+2f(k)=0; \quad f(0)=0, f(1)=1.$$

Solution. Taking z-transforms of both sides of the preceeding difference equation, we get

$$z^2F(z)-z^2f(0)-zf(1)+3zF(z)-3zf(0)+2F(z)=0$$

or, $$z^2F(z)-z+3zF(z)+2F(z)=0$$

or, $$(z^2+3z+2)F(z)=z$$

or, $$F(z)=\frac{z}{z^2+3z+2}$$

$$=\frac{z}{(z+1)(z+2)}$$

$$=\frac{z}{(z+2)}\Bigg|_{z=-1}\frac{1}{(z+1)}+\frac{z}{(z+1)}\Bigg|_{z=-2}\frac{1}{(z+2)}$$

$$=\frac{-1}{(z+1)}+\frac{2}{(z+2)}.$$

Taking the inverse z-transform, we get

$$f(k)=-1(-1)^{k-1}+2(-2)^{k-1;}k\geq 1$$

or, $$f(k)=(-1)^k-(-2)^k \qquad\qquad \textbf{Ans.}$$

Example 2.8. *Find the response of the system described by the difference equation*

$$f(k+2)-5f(k+1)+6f(k)=u(k)$$

given that, $f(0) = 0$ and $f(1) = 1$.

Solution. Taking z-transforms of both sides of the preceeding difference-equation, we get

$$z^2 F(z) - z^2 f(0) - zf(1) - 5zF(z) + 5zf(0) + 6F(z) = \mathcal{Z}[u(k)]$$

or,
$$z^2 F(z) - z - 5zF(z) + 6F(z) = \frac{z}{z-1}$$

or,
$$(z^2 - 5z + 6)F(z) = z + \frac{z}{z-1}$$

or,
$$(z^2 - 5z + 6)F(z) = \frac{z^2}{z-1}$$

or,
$$F(z) = \frac{z^2}{(z-1)(z^2 - 5z + 6)}$$

$$= \frac{z^2}{(z-1)(z-2)(z-3)}$$

$$= \frac{z^2}{(z-2)(z-3)}\bigg|_{z=1} \frac{1}{(z-1)} + \frac{z^2}{(z-1)(z-3)}\bigg|_{z=2} \frac{1}{(z-2)} + \frac{z^2}{(z-1)(z-2)}\bigg|_{z=3} \frac{1}{(z-3)}$$

$$= \frac{1}{2} \cdot \frac{1}{(z-1)} - \frac{4}{(z-2)} + \frac{9}{2} \cdot \frac{1}{(z-3)}.$$

Taking the inverse z-transform, we get

$$f(k) = \frac{1}{2}(1)^{k-1} - 4(2)^{k-1} + \frac{9}{2}(3)^{k-1}; k \geq 1$$

or,
$$f(k) = \frac{1}{2}(1)^k - 2(2)^k + \frac{3}{2}(3)^k \qquad \textbf{Ans.}$$

2.6   THE z-TRANSFER FUNCTION (PULSE TRANSFER FUNCTION)

Figure 2.10 shows a linear-discrete-time system wherein the output-sequence $y(k)$ of the system is given by the convolution sum between the input sequence $u(k)$ and the **impulse-response function** $g(k)$.

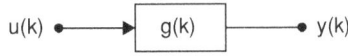

FIGURE 2.10   Linear discrete-time system.

Therefore, we may write

$$y(k) = \sum_{m=0}^{\infty} u(m)\, g(k-m). \qquad (2.32)$$

Taking the z-transform of the convolution sum (2.32), we have

$$Y(z) = \sum_{k=0}^{\infty} y(k)\, z^{-k}$$

or,
$$Y(z) = \sum_{k=0}^{\infty} \left[\sum_{m=0}^{\infty} u(m)\, g(k-m) \right] z^{-k}.$$

Interchanging the order of summation, we get

$$Y(z) = \sum_{m=0}^{\infty} u(m) \sum_{k=0}^{\infty} g(k-m)\, z^{-k}$$

or,
$$Y(z) = \sum_{m=0}^{\infty} u(m) z^{-m} \sum_{k=0}^{\infty} g(k)\, z^{-k} \qquad \text{(Property 2.14)}$$

or,
$$Y(z) = U(z).G(z). \qquad (2.33)$$

This is an important result wherein $G(z)$ is interpreted as the **z-transfer function** or **pulse transfer function** of the linear discrete-time system as shown in Figure 2.11.

Thus, from Equation (2.33), the pulse-transfer function is expressed as

$$G(z) = \frac{Y(z)}{U(z)}. \qquad (2.34)$$

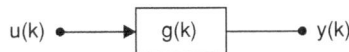

FIGURE 2.11   Block-diagram for a pulse-transfer function system.

It should be noted that $G(z)$ is the z-transform of the **impulse-response function** $g(k)$ of an initially relaxed system, i.e., a system with **zero initial conditions**.

As seen from Equation (2.34), we can define the z-transfer function (pulse transfer function) as the ratio of the z-transform of the output to the z-transform of the input, keeping all initial conditions zero, i.e., initially the system is relaxed.

$$G(z) = \frac{Y(z)}{U(z)} = \frac{z\text{-transform of output}}{z\text{-transform of input}}$$

Hence, the z-transfer function is a **mathematical model of an initially relaxed system**. From the knowledge of this model the output of the system to any input can be obtained by using the relation (2.34) and then taking the inverse z-transform to obtain $y(k)$. The z-transform function has the same important role in z-transform analysis of discrete-time systems as the transfer function in the Laplace-transform analysis of continuous-time systems.

2.6.1 Pulse Transfer-Function of Cascaded-Elements

2.6.1.1 Two Elements in Cascade, Each using a Sampler in the Input

Figure 2.12 shows two elements, having transfer functions $G_1(s)$ and $G_2(s)$ respectively, connected in cascade such that the input to each element is sampled and both the samplers S_1 and S_2 are operating in synchronism.

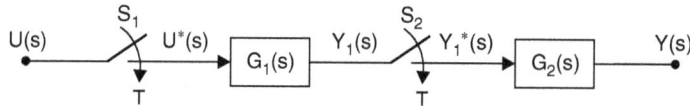

FIGURE 2.12   Two elements in cascade, each using a sampler in the input.

On observation, we may write

$$Y_1(s) = G_1(s) U^*(s)$$

and

$$Y(s) = G_2(s) Y_1^*(s).$$

Since

$$Y_1^*(s) = [G_1(s) U^*(s)]^* = G_1^*(s) U^*(s).$$

(As starring a starred function results in the starred function itself.)

Therefore,

$$Y(s) = G_2(s)G_1^*(s)U^*(s)$$

or, $$Y^*(s) = [G_2(s)G_1^*(s)\, U^*(s)]^*$$

or, $$Y^*(s) = G_2^*(s)G_1^*(s)\, U^*(s)$$

or, $$Y^*(s) = G_1^*(s)G_2^*(s)\, U^*(s). \tag{2.35}$$

In terms of a z-transform, Equation (2.35) can be written as

$$Y(z) = G_1(z)\, G_2(z)\, U(z).$$

$\therefore$ Pulse transfer-function is

$$\frac{Y(z)}{U(z)} = G_1(z)\, G_2(z). \tag{2.36}$$

The block diagram representation of pulse transfer function (2.36) in the z-domain is shown in Figure 2.13.

FIGURE 2.13   Block diagram representation of pulse transfer function (2.36).

2.6.1.2 Two Elements in Cascade, using Sampler in the Input of the First Element Only

Figure 2.14(a) shows two elements, having transfer functions $G_1(s)$ and $G_2(s)$ respectively, connected in cascade such that the input is given only to the first element sampled.

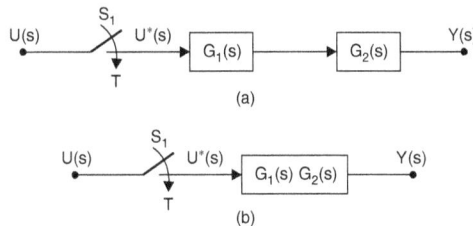

FIGURE 2.14   (a) Two elements in cascade, using the sampler in the input of the first element only; (b) After block-diagram reduction.

As per the block-diagram reduction process, Figure 2.14(a) is redrawn as shown in Figure 2.14(b).

On observation, we may write

$$Y(s) = G_2(s)G_1^*(s)U^*(s)$$

$\therefore$ $$Y^*(s) = [G_2(s)G_1^*(s)\,U^*(s)]^*$$

or, $$Y^*(s) = G_2^*(s)G_1^*(s)\,U^*(s)$$

or, $$Y^*(s) = G_1^*(s)G_2^*(s)\,U^*(s). \qquad (2.37)$$

Note that the term $G_1 G_2^*(s)$ indicates that the starring is done after combining $G_1(s)$ and $G_2(s)$ as per block-diagram reduction rules.

In terms of the z-transform, Equation (2.37) can be written as

$$Y(z) = G_1 G_2(z)U(z).$$

$\therefore$ Pulse transfer-function is

$$\frac{Y(z)}{U(z)} = G_1 G_2(z). \qquad (2.38)$$

The block-diagram representation of pulse-transfer function (2.38) in the z-domain is shown in Figure 2.15.

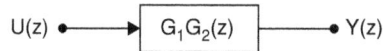

U(z) •——→ | $G_1 G_2(z)$ | ——→ • Y(z)

FIGURE 2.15   **Block-diagram representation of pulse-transfer function (2.38).**

Note that $G_1 G_2(z) \neq G_1(z)\,G_2(z)$, therefore, we must be careful and observe whether or not there is a sampler in between the cascaded elements. This is illustrated in Example (2.9).

Example 2.9. *Obtain the pulse-transfer function for the systems shown in Figure 2.12 and Figure 2.14 (a). Given that*

$$G_1(s) = \frac{1}{s+1}$$

and $$G_2(s) = \frac{1}{s+2}.$$

Solution. (*i*) Consider Figure 2.12.
We are given that

$$G_1(s) = \frac{1}{s+1} \text{ and } G_2(s) = \frac{1}{s+2}.$$

∴ $g_1(kT) = e^{-kT} \text{ and } g_2(kT) = e^{-2kT}.$

Taking the *z*-transforms, we have

$$G_1(z) = \frac{z}{z - e^{-T}} \text{ and } G_2(z) = \frac{z}{z - e^{-2T}}.$$

The pulse-transfer function is given by

$$\frac{Y(z)}{U(z)} = G_1(z)G_2(z) \qquad \text{[From Equation (2.36)]}$$

or, $$\frac{Y(z)}{U(z)} = \left[\frac{z}{z - e^{-T}}\right]\left[\frac{z}{z - e^{-2T}}\right]$$

or, $$\frac{Y(z)}{U(z)} = \frac{z^2}{(z - e^{-T})(z - e^{-2T})} \qquad\qquad \textbf{Ans.} \ (2.39)$$

(*ii*) Now, consider Figure 2.14 (*a*). As per the block diagram reduction process, it can be redrawn as shown in Figure 2.14(*b*).
We are given that

$$G_1(s) = \frac{1}{s+1} \text{ and } G_2(s) = \frac{1}{s+2}.$$

∴ $$G_1(s)G_2(s) = \frac{1}{(s+1)(s+2)} = \frac{1}{s+1} - \frac{1}{s+2}.$$

Therefore,

$$g_1 g_2(kT) = e^{-kT} - e^{-2kT}.$$

Taking the z-transforms, we have

$$G_1G_2(z) = \frac{z}{z - e^{-T}} - \frac{z}{z - e^{-2T}}$$

or,

$$G_1G_2(z) = \frac{z(e^{-T} - e^{-2T})}{(z - e^{-T})(z - e^{-2T})}.$$

The pulse transfer function is given by

$$\frac{Y(z)}{U(z)} = G_1G_2(z) = \frac{z(e^{-T} - e^{-2T})}{(z - e^{-T})(z - e^{-2T})} \qquad \textbf{Ans.} \quad (2.40)$$

On comparison of pulse transfer functions (2.39) and (2.40), it is concluded that

$$G_1G_2(z) \neq G_1(z)\, G_2(z). \tag{2.41}$$

2.6.2   Pulse Transfer Function of Closed-Loop Systems

2.6.2.1 Closed-Loop System, wherein the input signal is being sampled

Figure 2.16 (*a*) shows a closed-loop system having forward path transfer function $G(s)$ and feedback transfer function $H(s)$ and the input signal being sampled. As per the block-diagram reduction process, Figure 2.16(*a*) is redrawn as shown in Figure 2.16(*b*).

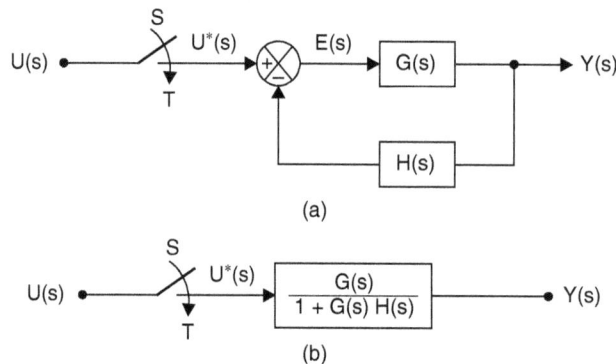

FIGURE 2.16   (*a*) Closed Loop system, wherein the input signal is being sampled;
(*b*) After block-diagram reduction.

On observation, we may write

$$Y(s) = \frac{G(s)}{1+G(s)H(s)} U^*(s).$$

$\therefore$
$$Y^*(s) = \left[\frac{G(s)}{1+G(s)H(s)} U^*(s) \right]^*$$

or,
$$Y^*(s) = \left[\frac{G(s)}{1+G(s)H(s)} \right]^* [U^*(s)]^*$$

or,
$$Y^*(s) = \left[\frac{G(s)}{1+G(s)H(s)} \right]^* U^*(s)$$

or,
$$Y^*(s) = \left[\frac{G(s)}{1+GH}(s) \right]^* U^*(s). \tag{2.42}$$

Note that the term $\left[\dfrac{G}{1+GH}(s) \right]^*$ indicates that the starring is done after combining $G(s)$ and $H(s)$ as per block-diagram reduction rules.

In terms of the z-transform, Equation (2.42) can be written as

$$Y(z) = \left[\frac{G}{1+GH}(z) \right] U(z).$$

$\therefore$ Pulse transfer function is

$$\frac{Y(z)}{U(z)} = \frac{G}{1+GH}(z). \tag{2.43}$$

The block-diagram representation of the pulse-transfer function is shown in Figure 2.17.

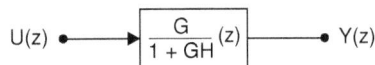

U(z) $\longrightarrow$ $\boxed{\dfrac{G}{1+GH}(z)}$ $\longrightarrow$ Y(z)

FIGURE 2.17   **Block-diagram representation of the pulse transfer function (2.43).**

2.6.2.2 Closed-Loop System, wherein the Error-Signal is being sampled

Figure 2.18 shows a closed-loop system having forward path transfer function $G(s)$ and feedback transfer function $H(s)$ and the error-signal being sampled.

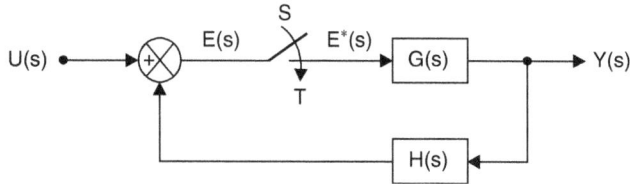

FIGURE 2.18   Closed-loop system, wherein the error-signal is sampled.

On observation, we may write

$$Y(s) = G(s)E^*(s)$$

and,
$$E(s) = U(s) - H(s)\,Y(s)$$

or,
$$E(s) = U(s) - G(s)\,H(s)\,E^*(s)$$

or,
$$E^*(s) = [U(s) - G(s)\,H(s)\,E^*(s)]^*$$

or,
$$E^*(s) = U^*(s) - [G(s)\,H(s)]^*\,E^*(s)$$

or,
$$E^*(s) = U^*(s) - GH^*(s)\,E^*(s)$$

or,
$$E^*(s)\,[1 + GH^*(s)] = U^*(s)$$

or,
$$E^*(s) = \frac{U^*(s)}{1 + GH^*(s)}. \tag{2.45}$$

Now, starring Equation (2.44), we have

$$Y^*(s) = [G(s)\,E^*(s)]^*$$

or,
$$Y^*(s) = G^*(s)\,E^*(s). \tag{2.46}$$

Substituting $E^*(s)$ from Equation (2.45) into (2.46), we get

$$Y^*(s) = \frac{G^*(s)\,U^*(s)}{1 + GH^*(s)}.$$

In terms of the z-transform, we have

$$Y(z) = \frac{G(z)\,U(z)}{1 + GH(z)}.$$

The pulse transfer function is given by

$$\frac{Y(z)}{U(z)} = \frac{G(z)}{1 + GH(z)}. \tag{2.47}$$

The block-diagram representation of pulse-transfer function (2.47) in the z-domain is shown in Figure 2.19.

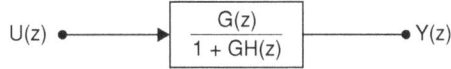

U(z) •————————→ | $\dfrac{G(z)}{1 + GH(z)}$ | ————————→• Y(z)

FIGURE 2.19   Block-diagram representation of pulse transfer function (2.47).

2.6.2.3 Closed-Loop System, wherein the Error-signal as well as the Feedback signal Are being Sampled

Figure 2.20 shows a closed-loop system having forward path transfer function $G(s)$, feedback transfer function $H(s)$, an error-signal as well as a feed-back signal being sampled with samplers S_1 and S_2 operating in synchronism.

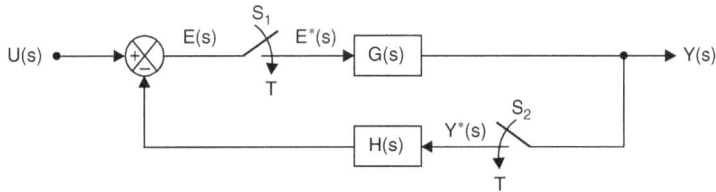

U(s) •——→⊗ —E(s)— S_1 —$E^*(s)$→ | G(s) | ——————————•——→ Y(s)
H(s) ←— S_2 ←— $Y^*(s)$

FIGURE 2.20   Closed-loop system, wherein the error-signal as well as the feedback signal are being sampled.

On observation, we may write

$$Y(s) = G(s)\,E^*(s) \tag{2.48}$$

and $E(s) = U(s) - H(s)\,Y^*(s)$

or, $E(s) = U(s) - H(s)\,[G(s)\,E^*(s)]^*$

or, $E(s) = U(s) - H(s)\,G^*(s)\,E^*(s)$

or,
$$E^*(s) = [U(s) - H(s) \, G^*(s) \, E^*(s)]^*$$

or,
$$E^*(s) = U^*(s) - G^*(s) \, H^*(s) \, E^*(s)$$

$$E^*(s)[1 + G^*(s) \, H^*(s)] = U^*(s)$$

or,
$$E^*(s) = \frac{U^*(s)}{1 + G^*(s) \, H^*(s)}. \tag{2.49}$$

Now, from Equation (2.48), we have

$$Y^*(s) = [G(s) \, E^*(s)]^*$$

or,
$$Y^*(s) = G^*(s) \, E^*(s). \tag{2.50}$$

Putting $E^*(s)$ from Equation (2.49) into (2.50), we get

$$Y^*(s) = \frac{G^*(s)U^*(s)}{1 + G^*(s)H^*(s)}.$$

In terms of the z-transform, we have

$$Y(z) = \frac{G(z) \, U(z)}{1 + G(z)H(z)}.$$

The pulse-transfer function is given by

$$\frac{Y(z)}{U(z)} = \frac{G(z)}{1 + G(z) \, H(z)}. \tag{2.51}$$

The block-diagram representation of pulse-transfer function (2.51) in the z-domain is shown in Figure 2.21.

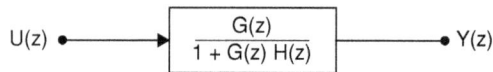

U(z) •———▶ | $\dfrac{G(z)}{1 + G(z)\,H(z)}$ | ———• Y(z)

FIGURE 2.21   Block-diagram representation of pulse-transfer function (2.51).

Example 2.10. *Obtain the pulse transfer function for the error-sampled system shown in Figure 2.18 given that*

$$G(s) = \frac{1}{s(s+1)}, \; H(s) = 1$$

and the sampling time being $T = 1$ sec.

Solution. The forward path transfer function

$$G(s) = \frac{1}{s(s+1)}$$

$$= \frac{1}{(s+1)}\bigg|_{s=0} \frac{1}{s} + \frac{1}{s}\bigg|_{s=-1} \frac{1}{(s+1)}$$

$$= \frac{1}{s} - \frac{1}{s+1}$$

$$\therefore \qquad g(kT) = u(kT) - e^{-kT}.$$

Taking the z-transform, we get

$$G(z) = \frac{z}{z-1} - \frac{z}{z - e^{-T}}.$$

Since, $T = 1$ sec.

$$\therefore \qquad G(z) = \frac{z}{z-1} - \frac{z}{z - e^{-1}}$$

$$= \frac{z}{z-1} - \frac{z}{z - 0.3679}$$

$$= \frac{0.6321z}{(z-1)(z-0.3679)}.$$

Now, the feedback transfer function

$$H(s) = 1$$

$$\therefore \qquad GH(s) = \frac{1}{s(s+1)} \times 1 = \frac{1}{s(s+1)} = \frac{1}{s} - \frac{1}{s+1}$$

$$GH(z) = \frac{z}{z-1} - \frac{z}{z - e^{-T}}.$$

Since $T = 1$ sec.

$$\therefore \qquad G(z) = \frac{z}{z-1} - \frac{z}{z - e^{-1}}$$

$$= \frac{z}{z-1} - \frac{z}{z - 0.3679}$$

$$= \frac{0.6321z}{(z-1)(z-0.3679)}.$$

Now, the Pulse transfer function for the error-sampled system is given by

$$\frac{Y(z)}{U(z)} = \frac{G(z)}{1 + GH(z)}$$

$$= \frac{\dfrac{0.6321z}{(z-1)(z-0.3679)}}{1 + \dfrac{0.6321z}{(z-1)(z-0.3679)}}$$

$$= \frac{0.6321z}{(z-1)(z-0.3679) + 0.6321z}$$

$$= \frac{0.6321z}{z^2 - 0.7358z + 0.3679} \qquad\qquad \textbf{Ans.}$$

2.7   STATE-SPACE REPRESENTATION OF DISCRETE-TIME SYSTEMS

The state-space approach for analysis and design of continuous-time systems can be extended for the analysis and design of discrete-time systems. The discrete form of the state-space representation is quite analogous to the continuous form. The general form of the state model for a **multivariable** discrete-time system is

$$\mathbf{x}(k+1) = \mathbf{f}(\mathbf{x}(k),\ \mathbf{u}(k));\qquad \text{State Equation} \qquad (2.52\ A)$$
$$\mathbf{y}(k) = \mathbf{g}(\mathbf{x}(k),\ \mathbf{u}(k));\qquad \text{Output Equation.} \quad (2.52\ B)$$

For a linear time-invariant system, we may write the Equations (2.52) as

$$\mathbf{x}(k+1) = \mathbf{A}\mathbf{x}(k) + \mathbf{B}\mathbf{u}(k);\qquad \text{State Equation} \qquad (2.52\ A)$$
$$\mathbf{y}(k) = \mathbf{C}\mathbf{x}(k) + \mathbf{D}\mathbf{u}(k);\qquad \text{Output Equation.} \quad (2.52\ B)$$

where, $\mathbf{x}(k) \to n \times 1$ State Vector,

$\mathbf{u}(k) \to m \times 1$ Input Vector,

$\mathbf{y}(k) \to p \times 1$ Output Vector,

$\mathbf{A} \to n \times n$ System Matrix,

$\mathbf{B} \to n \times m$ Input Matrix,

$\mathbf{C} \rightarrow p \times n$ Output Matrix,

$\mathbf{D} \rightarrow p \times m$ Transmission Matrix.

However, for a linear time-variant system, the state-model is given as

$$\mathbf{x}(k+1) = \mathbf{A}(k)\,\mathbf{x}(k) + \mathbf{B}(k)\,\mathbf{u}(k); \quad \text{State Equation} \quad (2.52\,A)$$

$$\mathbf{y}(k) = \mathbf{C}(k)\,\mathbf{x}(k) + \mathbf{D}(k)\,\mathbf{u}(k); \quad \text{Output Equation.} \ (2.52\,B)$$

In this chapter, we will be concerned mostly with systems described by Equations (2.53). The block-diagram representation of the state-model of linear multivariable time-invariant systems is shown in Figure (2.22).

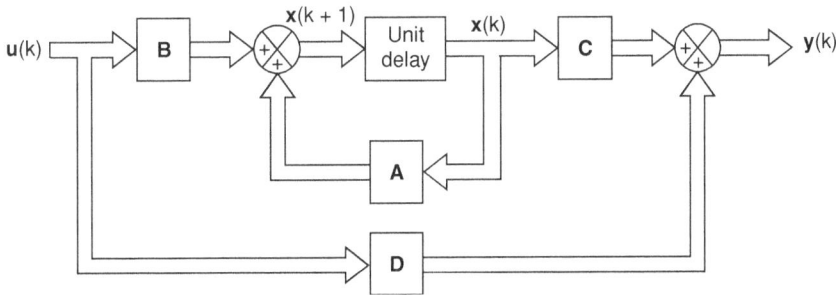

FIGURE 2.22   Block-diagram representation of the state-model of linear multivariable time-invariant systems.

2.8   SOLUTION OF STATE-EQUATIONS FOR LINEAR TIME-INVARIANT DISCRETE-TIME SYSTEMS

2.8.1   By Use of Recursion Procedure

Considering Equation (2.53A) we may write

$$\mathbf{x}(1) = \mathbf{A}\mathbf{x}(0) + \mathbf{B}\mathbf{u}(0)$$

$$\mathbf{x}(2) = \mathbf{A}\mathbf{x}(1) + \mathbf{B}\mathbf{u}(1)$$

$$= \mathbf{A}\{\mathbf{A}\mathbf{x}(0) + \mathbf{B}\mathbf{u}(0)\} + \mathbf{B}\mathbf{u}(1)$$

$$= \mathbf{A}^2\mathbf{x}(0) + \mathbf{A}\mathbf{B}\mathbf{u}(0) + \mathbf{B}\mathbf{u}(1)$$

$$\vdots \qquad \vdots \qquad \vdots$$

$$\mathbf{x}(k) = \mathbf{A}^k\mathbf{x}(0) + \mathbf{A}^{k-1}\mathbf{B}\mathbf{u}(0) + \mathbf{A}^{k-2}\mathbf{B}\mathbf{u}(1) + \ldots\ldots + \mathbf{B}\mathbf{u}(k-1)$$

or, we may write

$$\mathbf{x}(k) = \mathbf{A}^k \mathbf{x}(0) + \sum_{i=0}^{k-1} \mathbf{A}^{k-i-1} \mathbf{Bu}(i). \tag{2.55}$$

Equation (2.55) gives the solution of state-equation (2.53A).

2.8.2 z-Transform Approach to the Solution of Discrete-Time State Equations

Consider Equation (2.53A). Taking the z-transform on both sides of Equation (2.53A), we have

$$z\mathbf{X}(z) - z\mathbf{x}(0) = \mathbf{AX}(z) + \mathbf{BU}(z)$$

or, $$(z\mathbf{I} - \mathbf{A})\mathbf{X}(z) = z\mathbf{x}(0) + \mathbf{BU}(z)$$

or, $$\mathbf{X}(z) = (z\mathbf{I} - \mathbf{A})^{-1} z\mathbf{x}(0) + (z\mathbf{I} - \mathbf{A})^{-1} \mathbf{BU}(z). \tag{2.56}$$

Taking the inverse z-transform on both sides of Equation (2.56), we have

$$\mathbf{x}(k) = \mathcal{Z}^{-1}\{(z\mathbf{I} - \mathbf{A})^{-1} z\} \mathbf{x}(0) + \mathcal{Z}^{-1}\{(z\mathbf{I} - \mathbf{A})^{-1} \mathbf{BU}(z)\} \tag{2.57}$$

Equation (2.57) also gives the solution of state Equation (2.53A).

Now, comparing Equations (2.55) and (2.57), we have two important results, i.e.,

$$\mathbf{A}^k = Z^{-1}\{(z\mathbf{I} - \mathbf{A})^{-1} z\} \tag{2.58}$$

and, $$\sum_{i=0}^{k-1} \mathbf{A}^{k-i-1}\mathbf{Bu}(i) = Z^{-1}\{(z\mathbf{I} - \mathbf{A})^{-1} \mathbf{BU}(z)\}. \tag{2.59}$$

The solution of state Equation (2.53A) given by (2.55) may also be rewritten as

$$\mathbf{x}(k) = \phi(k) \mathbf{x}(0) + \sum_{i=0}^{k-1} \phi(k - i - 1)\mathbf{Bu}(i) \tag{2.60}$$

where, $\phi(k) = \mathbf{A}^k$ and is referred to as the **State Transition Matrix** for the discrete-time system described by state-model (2.53). The **properties** of state transition matrix, $\phi(k)$ are summarized here:

1. $\phi(0) = I$
2. $\phi^{-1}(k) = \phi(-k)$
3. $\phi(k, k_0) = \phi(k - k_0) = A^{(k-k_0)}; \ k > k_0$

Example 2.11. *Determine the solution of the discrete-time state equation,*

$$\mathbf{x}(k+1) = \begin{bmatrix} 0 & 1 \\ -0.16 & -1 \end{bmatrix} \mathbf{x}(k) + \begin{bmatrix} 1 \\ 1 \end{bmatrix} u(k)$$

where, $u(k) = 1$ for $k = 0, 1, 2, \ldots$. Assume that initial conditions are

$$\mathbf{x}(0) = \begin{bmatrix} x_1(0) \\ x_2(0) \end{bmatrix} = \begin{bmatrix} 1 \\ -1 \end{bmatrix}.$$

Solution.

Here,
$$\mathbf{A} = \begin{bmatrix} 0 & 1 \\ -0.16 & -1 \end{bmatrix}.$$

$\therefore$
$$[z\mathbf{I} - \mathbf{A}] = \begin{bmatrix} z & -1 \\ 0.16 & z+1 \end{bmatrix}.$$

Now,
$$\text{Adj } [z\mathbf{I} - \mathbf{A}] = \begin{bmatrix} z+1 & -0.16 \\ 1 & z \end{bmatrix}^T$$

$$= \begin{bmatrix} z+1 & 1 \\ -0.16 & z \end{bmatrix}$$

and,
$$|z\mathbf{I} - \mathbf{A}| = z(z+1) + 0.16$$
$$= z^2 + z + 0.16$$
$$= (z + 0.2)(z + 0.8).$$

$\therefore$
$$[z\mathbf{I} - \mathbf{A}]^{-1} = \frac{\text{Adj}[z\mathbf{I} - \mathbf{A}]}{|z\mathbf{I} - \mathbf{A}|}$$

$$= \begin{bmatrix} \dfrac{z+1}{(z+0.2)(z+0.8)} & \dfrac{1}{(z+0.2)(z+0.8)} \\[2ex] \dfrac{-0.16}{(z+0.2)(z+0.8)} & \dfrac{z}{(z+0.2)(z+0.8)} \end{bmatrix}.$$

Now,

$$\mathbf{X}(z) = (z\mathbf{I} - \mathbf{A})^{-1}z\mathbf{x}(0) + (z\mathbf{I} - \mathbf{A})^{-1}\mathbf{Bu}(z)$$
$$= (z\mathbf{I} - \mathbf{A})^{-1}[z\mathbf{x}(0) + \mathbf{Bu}(z)].$$

$$\because \qquad \mathbf{U}(z) = \frac{z}{z-1}.$$

$$\therefore \qquad z\mathbf{x}(0) + \mathbf{BU}(z) = \begin{bmatrix} z \\ -z \end{bmatrix} + \begin{bmatrix} \dfrac{z}{z-1} \\ \dfrac{z}{z-1} \end{bmatrix}$$

$$= \begin{bmatrix} \dfrac{z^2}{z-1} \\ \dfrac{-z^2 + 2z}{z-1} \end{bmatrix}.$$

Thus,

$$\mathbf{X}(z) = \begin{bmatrix} \dfrac{z+1}{(z+0.2)(z+0.8)} & \dfrac{1}{(z+0.2)(z+0.8)} \\ \dfrac{-0.16}{(z+0.2)(z+0.8)} & \dfrac{z}{(z+0.2)(z+0.8)} \end{bmatrix} \begin{bmatrix} \dfrac{z^2}{z-1} \\ \dfrac{-z^2 + 2z}{z-1} \end{bmatrix}$$

Or,

$$\mathbf{X}(z) = \begin{bmatrix} \dfrac{(z^2 + 2)z}{(z+0.2)(z+0.8)(z-1)} \\ \dfrac{(-z^2 + 1.84z)z}{(z+0.2)(z+0.8)(z-1)} \end{bmatrix}$$

$$= \begin{bmatrix} \left(\dfrac{-\dfrac{17}{6}z}{z+0.2} + \dfrac{\dfrac{22}{9}z}{z+0.8} + \dfrac{\dfrac{25}{18}z}{z-1} \right) \\ \left(\dfrac{\dfrac{3.4}{6}z}{z+0.2} + \dfrac{-\dfrac{17.6}{9}z}{z+0.8} + \dfrac{\dfrac{7}{18}z}{z-1} \right) \end{bmatrix}.$$

Therefore,

$$\mathbf{x}(k) = \mathcal{Z}^{-1}\mathbf{X}(z) = \begin{bmatrix} -\dfrac{17}{6}(-0.2)^k + \dfrac{22}{9}(-0.8)^k + \dfrac{25}{18} \\[4mm] \dfrac{3.4}{6}(-0.2)^k - \dfrac{17.6}{9}(-0.8)^k + \dfrac{7}{18} \end{bmatrix}$$

Ans.

EXERCISES

1. Determine the z-transform of the following:

   (a) $f(k) = k^2$
   (b) $f(k) = k^2 a^{k-1}$
   (c) $f(k) = \sinh \beta k$
   (d) $f(k) = \cosh \beta k$

2. Find the z-transform of the following continuous-time functions:

   (a) e^{at}
   (b) te^{-at}
   (c) $e^{-at} \sin \omega t$
   (d) $2u(t) + t$

3. Find the impulse train Laplace transform (starred transform) $F^*(s)$ for the following continuous time functions:

   (a) $f(t) = u(t)$
   (b) $f(t) = e^{-at}$

4. Find the inverse z-transform of the following:

   (a) $F(z) = \dfrac{z}{(z-1)(z+0.85)}$

   (b) $F(z) = \dfrac{z^2}{(z-1)(z^2-5z+6)}$

   (c) $F(z) = \dfrac{z^2}{(z-1)(z^2-1.6z+1.64)}$

5. Find the z-domain transfer function of the following s-domain transfer functions:

   (a) $\dfrac{\omega}{s^2 + \omega^2}$

   (b) $\dfrac{\omega}{(s+\alpha)^2 + \omega^2}$

   (c) $\dfrac{s}{s^2 + \omega^2}$

   (d) $\dfrac{s+\alpha}{(s+\alpha)^2 + \omega^2}$.

6. Determine the z-transfer function of two cascaded systems each described by the difference equation

$$y(k) = 0.5\, y(k-1) + u(k).$$

7. Solve the following difference equation by means of the z-transform

$$f(k+2) - f(k) = 0; f(0) = 1, f(1) = 0.$$

8. Solve the difference equation

$$y(k+2) + 3y(k+1) + 2y(k) = u(k); y(0) = 1$$
$$y(k) = 0 \text{ for } k < 0.$$

   [**Hint:** $y(1)$ is needed in the solution, which can be obtained by putting k = - 1 in the preceeding difference equation, viz.

$$y(1) + 3y(0) + 2y(-1) = u(-1)$$
or,
$$y(1) = -3.]$$

9. Determine the unit-step time response for the pulse transfer function

$$\frac{Y(z)}{U(z)} = \frac{z}{z^2 - z + 0.5}.$$

10. Determine the pulse transfer function for the sampled data control system shown in Figure 2.23.

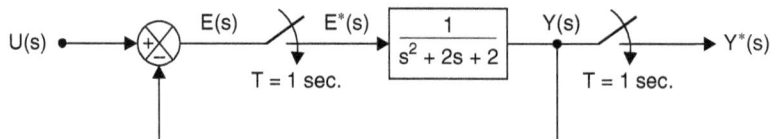

FIGURE 2.23

11. A discrete-time system is described by the difference equation

$$y(k+2) + 5y(k+1) + 6y(k) = u(k)$$

$$y(0) = y(1) = 0; T = 1 \text{ sec.}$$

(a) Determine the state-model in normal form.
(b) Compute STM.
(c) For input u(k) = 1; k $\geq$ 0 find the output y(k).

REFERENCES

(1) Benjamin C. KuO, *Digital Control Systems*, Holt, Rinehart and Winston, Inc., 1980.

(2) B. S. Manke, *Linear Control Systems*, Khanna Publishers, Delhi, 6th Edition, 2003.

(3) Constantine H. Houpis, Gary B. Lamont, *Digital Control Systems*, McGraw-Hill, Inc., Singapore, 1985.

(4) D. Roy Choudhary, *Modern Control Engineering*, Prentice-Hall of India Pvt. Ltd., New Delhi, 2005.

(5) Francis H. Raven, *Automatic Control Engineering*, McGraw-Hill, Inc., Singapore, 4th Edition, 1987.

(6) Gene F. Franklin, J. David Powell, Michael Workman, *Digital Control of Dynamic Systems*, Pearson Education , Inc., Singapore, 3rd Edition, 2002.

(7) I.J. Nagrath, M. Gopal, *Control Systems Engineering*, New Age International Publishers, New Delhi, 4th Edition, 2005.

3

STABILITY ANALYSIS OF NONLINEAR SYSTEMS

3.1   INTRODUCTION

The stability of a system implies that the small changes in the system input (either in system parameters or in initial conditions of the system) do not result in large changes in the system output. For a given control system, stability is one of the most important characteristics to be determined. In order to analyze the stability of linear time-invariant systems, many stability criteria, such as Nyquist stability criterion, Routh's stability criterion, etc., are available. However, for nonlinear systems and/or time-varying systems, such stability criteria do not apply.

In 1892, **A.M. Lyapunov**, the Russian mathematician, presented two methods (known as first and second methods of Lyapunov) in order to determine the stability of dynamic systems described by ordinary differential equations. By using the **second method of Lyapunov**, the stability of the dynamic system can be determined without actually solving the differential equations, that's why it is also referred to as the **direct method of Lyapunov**. This is quite advantageous because solving nonlinear and/or time-varying state equations is usually very difficult or sometimes may be impossible, too. Thus, the direct-method of Lyapunov is the most general method that can be employed for the determination of the stability of nonlinear and/or time-varying systems as it avoids the necessity of solving state equations.

3.2   AUTONOMOUS SYSTEM AND EQUILIBRIUM STATE

Consider the general state equation for a time-invariant system

$$\mathbf{x}(t) = \mathbf{f}(x(t), \mathbf{u}(t)). \tag{3.1}$$

If the input vector $\mathbf{u}(t)$ is constant, we may write Equation (3.1) as

$$\mathbf{x}(t) = \mathbf{F}(\mathbf{x}(t)). \tag{3.2}$$

The system described by Equation (3.2) is called an **autonomous system**.

In the system described by Equation (3.2), a state $\mathbf{x}_e(t)$ where

$$\mathbf{F}(\mathbf{x}_e(t)) = \mathbf{0}, \text{ for all } t$$

is called an **equilibrium state** of the system.

For studying the system dynamic response at an equilibrium point to small perturbation, the system is linearized at that point using the linearization technique. (Refer to Section 1.7.)

The linearized state model of the system described by Equation (3.2) may be written as

$$\mathbf{x}(t) = \mathbf{A}\mathbf{x}(t). \tag{3.3}$$

For this linear autonomous system, there exists only one equilibrium state $\mathbf{x}_e(t) = \mathbf{0}$ if $\mathbf{A}$ is nonsingular i.e., $|\mathbf{A}| \neq 0$, and there exist infinitely many equilibrium states if $\mathbf{A}$ is singular i.e., $|\mathbf{A}| = 0$. For the nonlinear systems, there may be one or more equilibrium states.

If $\mathbf{x}_e(t)$ is an isolated equilibrium state, it can be shifted to the origin of the state-space by a translation of coordinates. In this chapter, we shall deal with the stability analysis of only such equilibrium states.

3.3   STABILITY DEFINITIONS

As seen in Section 3.2, there may be one or more equilibrium states for the nonlinear systems. Thus, in the case of nonlinear systems, we shall define the system stability relative to the equilibrium state rather than using a general term '**stability of a system**.' There are a number of stability definitions in control systems' literature. We shall only concentrate on three of these: **stability**, **asymptotic stability**, and **asymptotic stability in-the-large**.

Consider an autonomous system described by the state Equation (3.2). We assume that the system has only one equilibrium state, which is the case with all **properly designed** systems. Moreover, without loss of generality, we assume the **origin** of state-space as the equilibrium point.

3.3.1 Stability in the Sense of Lyapunov

The system described by Equation (3.2) is said to be **stable at the origin** if, for every real number $\varepsilon > 0$, there exists a real number $\delta(\varepsilon) > 0$ such that $\|\mathbf{x}(t_0)\| \leq \delta$ results in $\|\mathbf{x}(t)\| \leq \varepsilon$ for all $t \geq t_0$.

In the preceeding definition, $\|\mathbf{x}(t)\|$ is called the **Euclidean norm** and is defined as

$$\|\mathbf{X}(t)\| = [x_1^{\,2}(t) + x_2^{\,2}(t) + \ldots\ldots + x_n^{\,2}(t)]^{1/2}.$$

Also, $\|\mathbf{x}(t)\| \leq R$ represents a **hyper-spherical region** $S(R)$ of radius R surrounding the equilibrium point $\mathbf{x}_e(t) = \mathbf{0}$.

Thus, the system described by Equation (3.2) is said to be stable at the origin if, corresponding to each $S(\varepsilon)$, there is an $S(\delta)$ such that trajectories starting in $S(\delta)$ do not leave $S(\varepsilon)$ as t increases indefinitely, as illustrated in Figure 3.1(*a*).

3.3.2 Asymptotic Stability in the Sense of Lyapunov

The system described by Equation (3.2) is said to be **asymptotically stable at the origin** if it is stable in the sense of Lyapunov and δ can be found such that $\|\mathbf{x}(t_0)\| \leq \delta$ results in $\mathbf{x}(t) \to 0$ as $t \to \infty$ i.e., trajectories starting within $S(\delta)$ converge on the origin without leaving $S(\varepsilon)$ as t increases indefinitely, as illustrated in Figure 3.1(*b*).

3.3.3 Asymptotic Stability in-the-Large

The system described by Equation (3.2) is said to be asymptotically stable in-the-large (**globally asymptotically stable**) at the origin if it is stable in the sense of Lyapunov and every initial state $\mathbf{x}(t_0)$, regardless of how near or far it is from the origin, results in $\mathbf{x}(t) \to \mathbf{0}$ as $t \to \infty$.

3.3.4 Instability

The system described by Equation (3.2) is said to be **unstable** if for some real number $\varepsilon > 0$ and any real number $\delta(\varepsilon) > 0$, no matter how small, there is always a state $\mathbf{x}(t_0)$ in $S(\delta)$ such that the trajectory starting at this state leaves $S(\varepsilon)$ as illustrated in Figure 3.1(c).

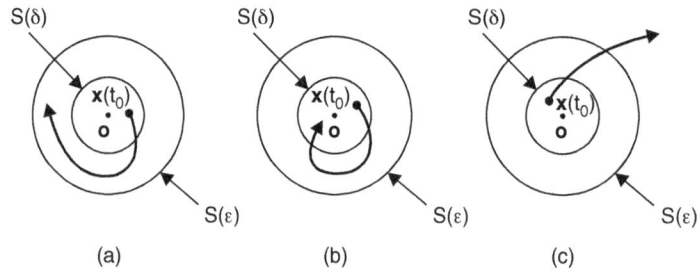

FIGURE 3.1   *(a)* **Stability in the sense of Lyapunov**
(b) **Asymptotic stability**
(c) **Instability**

3.4   CONCEPT OF SIGN-DEFINITENESS

In this section, we will discuss the concept of sign-definiteness of a scalar function V of a vector $\mathbf{x}$ i.e., $V(\mathbf{x})$.

3.4.1   Positive-Definiteness of a Scalar Function

A scalar function $V(\mathbf{x})$, which for some real number $\varepsilon > 0$, satisfying the following properties for all $\mathbf{x}$ in the region $\|\mathbf{x}\| \leq \varepsilon$

$$(a)\ V(\mathbf{x}) > 0 \ ; \mathbf{x} \neq 0$$
$$(b)\ V(0) = 0$$

is called a **positive-definite scalar function**.

For example, $V(\mathbf{x}) = x_1^2 + 2x_2^2$ is a positive-definite scalar function.

3.4.2   Negative-Definiteness of a Scalar Function

A scalar function $V(\mathbf{x})$, which for some real number $\varepsilon > 0$, satisfying the following properties, for all $\mathbf{x}$ in the region $\|\mathbf{x}\| \leq \varepsilon$

$$(a)\ V(\mathbf{x}) < 0\ ;\ \mathbf{x} \neq 0$$
$$(b)\ V(0) = 0$$

is called a **negative-definite scalar function**.

For example, $V(\mathbf{x}) = -x_1^2 - (x_1 + x_2)^2$ is a negative-definite scalar function.

3.4.3   Positive-Semidefiniteness of a Scalar Function

A scalar function $V(\mathbf{x})$, which for some real number $\varepsilon > 0$, satisfying the following properties for all $\mathbf{x}$ in the region $\|\mathbf{x}\| \leq \varepsilon$

$$(a)\ V(\mathbf{x}) \geq 0;\ \mathbf{x} \neq 0$$
$$(b)\ V(0) = 0$$

is called a **positive-semidefinite scalar function**.

For example, $V(\mathbf{x}) = (x_1 + x_2)^2$ is a positive-semidefinite scalar function.

3.4.4   Negative-Semidefiniteness of a Scalar Function

A scalar function $V(\mathbf{x})$, which for some real number $\varepsilon > 0$, satisfying the following properties for all $\mathbf{x}$ in the region $\|\mathbf{x}\| \leq \varepsilon$

$$(a)\ V(\mathbf{x}) \leq 0\ ;\ \mathbf{x} \neq 0$$
$$(b)\ V(0) = 0$$

is called a **negative-semidefinite scalar function**.

For example, $V(\mathbf{x}) = -(x_1 + x_2)^2$ is a negative-semidefinite scalar function.

It should be **noted** that in the definitions given in Sections 3.4.1, 3.4.2, 3.4.3, and 3.4.4, **if** ε is choosen arbitrarily large, the definitions hold in the entire state-space and are said to be **global**.

3.4.5 Indefiniteness of a Scalar function

A scalar function $V(\mathbf{x})$, which for some real number $\varepsilon > 0$, it assumes both positive and negative values, for all $\mathbf{x}$ within the region $\|\mathbf{x}\| \le \varepsilon$; no matter how small ε is choosen, is called an **indefinite scalar function**.
For example, $V(\mathbf{x}) = x_1 x_2 + x_2^2$ is an indefinite scalar function.

3.5   QUADRATIC FORM OF A SCALAR FUNCTION

The quadratic form of a scalar function plays an important role in the stability analysis based on Lyapunov's direct (second) method.
If a scalar function $V(\mathbf{x})$ is of the quadratic form, then it can be expressed as

$$V(\mathbf{x}) = \mathbf{x}^T \mathbf{P} \mathbf{x} \qquad (3.4)$$

where $\mathbf{x}$ is a real vector and $\mathbf{P}$ is a real symmetric matrix.
In the expanded form, Equation (3.4) may be rewritten as

$$V(\mathbf{x}) = [x_1 \ x_2 \ \dots \ x_n] \begin{bmatrix} p_{11} & p_{12} & \cdots & p_{1n} \\ p_{21} & p_{22} & \cdots & p_{2n} \\ & & & \\ p_{n1} & p_{n2} & \cdots & p_{nn} \end{bmatrix} \begin{bmatrix} x_1 \\ x_2 \\ \\ x_n \end{bmatrix}. \qquad (3.5)$$

3.6   DEFINITENESS OF A MATRIX (SYLVESTOR'S THEOREM)

If a scalar function $V(\mathbf{x})$ is of the quadratic form given by Equation (3.4), then the definiteness of $V(\mathbf{x})$ is attributed to matrix $\mathbf{P}$. The definiteness of matrix $\mathbf{P}$ can be determined by **Sylvestor's theorem**, which states that the **necessary and sufficient conditions** for a matrix

$$\mathbf{P} = \begin{bmatrix} p_{11} & \mathrm{p}_{12} & \cdots & p_{1n} \\ p_{21} & \mathrm{p}_{22} & \cdots & p_{2n} \\ & & & \\ p_{n1} & p_{n2} & \cdots & p_{nn} \end{bmatrix}$$

to be **positive-definite** are that all the **successive principal minors** of **P** be positive, i.e.,

$$p_{11} > 0$$

$$\begin{vmatrix} p_{11} & p_{12} \\ p_{21} & p_{22} \end{vmatrix} > 0$$

$$\begin{vmatrix} p_{11} & p_{12} & p_{13} \\ p_{21} & p_{22} & p_{23} \\ p_{31} & p_{32} & p_{33} \end{vmatrix} > 0$$

$$\begin{vmatrix} p_{11} & p_{12} & \cdots & p_{1n} \\ p_{21} & p_{22} & \cdots & p_{2n} \\ & & & \\ p_{n1} & p_{n2} & \cdots & p_{nn} \end{vmatrix} > 0.$$

Note that:

- The matrix **P** is positive-definite if all eigen-values of **P** are positive.
- The matrix **P** is positive-semidefinite if any of the principal minors of **P** are zero.
- The matrix **P** is negative-definite if the matrix $-$ **P** is positive-definite.
- The matrix **P** is negative-semidefinite if the matrix $-$ **P** is positive-semidefinite.
- If **P** is positive-definite, so is $\mathbf{P}^2$ and $\mathbf{P}^{-1}$.

It should be noted that the definiteness of a scalar function $V(\mathbf{x})$, which is of the quadratic form given by Equation (3.4), is **global**.

Example 3.1. *Determine whether the following quadratic form is positive-definite:*

$$V(\mathbf{x}) = 10x_1^{\,2} + 4x_2^{\,2} + x_3^{\,2} + 2x_1x_2 - 2x_2x_3 - 4x_1x_3 .$$

Solution. The quadratic form of scalar function $V(\mathbf{x})$ can be written as

$$V(\mathbf{x}) = \mathbf{x}^T \mathbf{P} \mathbf{x}. \tag{3.6}$$

The matrix $\mathbf{P}$ can be found as,

$$\mathbf{P} = \begin{bmatrix} \text{Coeff.}(x_1^2) & \frac{1}{2}\text{Coeff.}(x_1 x_2) & \frac{1}{2}\text{Coeff.}(x_1 x_3) \\ \frac{1}{2}\text{Coeff.}(x_2 x_1) & \text{Coeff.}(x_2^2) & \frac{1}{2}\text{Coeff.}(x_2 x_3) \\ \frac{1}{2}\text{Coeff.}(x_3 x_1) & \frac{1}{2}\text{Coeff.}(x_3 x_2) & \text{Coeff.}(x_3^2) \end{bmatrix}.$$

We may write Equation (3.6) as

$$V(\mathbf{x}) = \begin{bmatrix} x_1 & x_2 & x_3 \end{bmatrix} = \begin{bmatrix} 10 & 1 & -2 \\ 1 & 4 & -1 \\ -2 & -1 & 1 \end{bmatrix} \begin{bmatrix} x_1 \\ x_2 \\ x_3 \end{bmatrix}.$$

Applying Sylvestor's Theorem, we obtain

$$p_{11} = 10 > 0$$

$$\begin{vmatrix} p_{11} & p_{12} \\ p_{21} & p_{22} \end{vmatrix} = \begin{vmatrix} 10 & 1 \\ 1 & 4 \end{vmatrix} = 39 > 0$$

$$\begin{vmatrix} p_{11} & p_{12} & p_{13} \\ p_{21} & p_{22} & p_{23} \\ p_{31} & p_{32} & p_{33} \end{vmatrix} = \begin{vmatrix} 10 & 1 & -2 \\ 1 & 4 & -1 \\ -2 & -1 & 1 \end{vmatrix} = 17 > 0.$$

Since all the successive principal minors of the matrix $\mathbf{P}$ are positive, it means that $\mathbf{P}$ is positive-definite. Hence $V(\mathbf{x})$ is positive-definite.   **Ans.**

3.7   LYAPUNOV'S STABILITY CRITERION (DIRECT METHOD OF LYAPUNOV)

The direct method of Lyapunov is based on the **concept of energy** and the relation of stored energy with the system stability that the energy stored in a stable system cannot increase with time. Consider an autonomous physical system described as

$$\mathbf{x}(t) = \mathbf{F}(\mathbf{x}(t)).$$

Let $V(\mathbf{x})$ be the suitable function of energy associated with the system. If the derivative $\dfrac{dV(\mathbf{x})}{dt}$ is negative for all $\mathbf{x}$ except the equilibrium point $\mathbf{x}_e$, then it follows that energy associated with the system decays with increasing time t untill it finally assumes its minimum value at equilibrium point $\mathbf{x}_e$. This **holds good** because energy associated with a system being a **non-negative function of system state** reaches a **minimum** only if the system motion stops.

For purely mathematical systems, however, there is no obvious way of defining an "**energy-function**." In order to circumvent this difficulty, A.M. Lyapunov, the Russian mathematician, introduced a **fictitous energy function**, which is now known as **Lyapunov's function**. This idea is, however, more general than that of energy and is more widely applicable.

The method for investigating the stability of a system using Lyapunov's function (known as Lyapunov's direct method) is given by the following theorems:

Theorem 3.1. Consider the system described by

$$\mathbf{x}(t) = \mathbf{F}(\mathbf{x}(t)); \mathbf{F}(\mathbf{0}) = \mathbf{0}. \tag{3.7}$$

If there exists a scalar function $V(\mathbf{x})$, which for some real number $\varepsilon > 0$ satisfies the following properties for all $\mathbf{x}$ in the region $\|\mathbf{x}\| \leq \varepsilon$

(*a*) $V(\mathbf{x})$ is a positive-definite scalar function, i.e.,

$$V(\mathbf{x}) > 0; \mathbf{x} \neq \mathbf{0}$$

and, $V(\mathbf{0}) = 0,$

(*b*) $V(\mathbf{x})$ has continuous first partial derivatives with respect to all components of $\mathbf{x}$,

(c) $\dfrac{dV(\mathbf{x})}{dt}$ is a negative-semidefinie scalar function, i.e.,

$$\frac{dV(\mathbf{x})}{dt} \leq 0; \mathbf{x} \neq \mathbf{0}.$$

then the system described by (3.7) is **stable** at the origin.

Theorem 3.2. Consider the system described by

$$x(t) = \mathbf{F}(\mathbf{x}(t)); \mathbf{F}(\mathbf{0}) = 0. \qquad (3.8)$$

If there exists a scalar function $V(\mathbf{x})$, which for some real number $\varepsilon > 0$ satisfies the following properties for all $\mathbf{x}$ in the region $\|\mathbf{x}\| \leq \varepsilon$

(a) $V(\mathbf{x})$ is a positive-definite scalar function, i.e.,

$$V(\mathbf{x}) > 0; \mathbf{x} \neq \mathbf{0}$$

and, $$V(\mathbf{0}) = 0,$$

(b) $V(\mathbf{x})$ has continuous first partial derivatives with respect to all components of $\mathbf{x}$,

(c) $\dfrac{dV(\mathbf{x})}{dt}$ is a negative-definite scalar function, i.e.,

$$\frac{dV(\mathbf{x})}{dt} < 0; \mathbf{x} \neq \mathbf{0},$$

then the system described by (3.8) is **asymptotically stable** at the origin.

Theorem 3.3. Consider the system described by

$$x(t) = \mathbf{F}(\mathbf{x}(t)); \mathbf{F}(\mathbf{0}) = 0. \qquad (3.9)$$

If there exists a scalar function $V(\mathbf{x})$, which for some real number $\varepsilon > 0$ satisfies the following properties, for all $\mathbf{x}$ in the region $\|\mathbf{x}\| \leq \varepsilon$

(a) $V(\mathbf{x})$ is a positive definite scalar function, i.e.,

$$V(\mathbf{x}) > 0; \mathbf{x} \neq \mathbf{0}$$

and, $$V(\mathbf{0}) = 0,$$

(b) $V(\mathbf{x})$ has continuous first partial derivatives with respect to all components of $\mathbf{x}$,

(c) $\dfrac{dV(\mathbf{x})}{dt}$ is a negative-definite salar function, i.e.,

$$\frac{dV(x)}{dt} < 0; \mathbf{x} \neq \mathbf{0},$$

(d) $V(\mathbf{x}) \rightarrow \infty$ as $\|\mathbf{x}\| \rightarrow \infty$, then the system described by (3.9) is **asymptotically stable in-the-large** at the origin.

The determination of stability via Lyapunov's direct method centers around the choice of a positive-definite function $V(\mathbf{x})$ called the Lyapunov's function. Unfortunately, there is no obvious method for selecting a Lyapunov function that is unique for a special problem. The choice of a suitable **Lyapunov function depends on the ingenuity of the designer**. Note that if a suitable Lyapunov function can not be found, it no way implies that the system is unstable. It only means that our **attempt** in trying to establish the stability of the system has failed. The basic **instability theorem** can be stated as follows:

Theorem 3.4. Consider the system described by

$$\mathbf{x}(t) = \mathbf{F}(\mathbf{x}(t)); \ \mathbf{F}(\mathbf{0}) = 0. \tag{3.10}$$

If there exists a scalar function $V(\mathbf{x})$, which for some real number $\varepsilon > 0$ satisfies the following properties for all $\mathbf{x}$ in the region $\|\mathbf{x}\| \le \varepsilon$

(*a*) $V(\mathbf{x})$ is a positive-definite scalar function, i.e.,

$$V(\mathbf{x}) > 0; \mathbf{x} \neq \mathbf{0}$$

and,
$$V(0) = 0,$$

(*b*) $V(\mathbf{x})$ has continuous first partial derivatives with respect to all components of $\mathbf{x}$,

(c) $\dfrac{dV(\mathbf{x})}{dt}$ is a positive-definite (or semidefinite) scalar function, i.e.,

$$\frac{dV(\mathbf{x})}{dt} \ge 0; \mathbf{x} \neq \mathbf{0},$$

then the system described by (3.10) is unstable at the origin.

Example 3.2. *Consider a nonlinear system described by*

$$x_1 = x_2$$
$$x_2 = -x_1 - x_2^{\,3}.$$

Investigate the system's stability using the direct method of Lyapunov.

Solution. We might choose a positive-definite scalar function as,

$$V(\mathbf{x}) = x_1{}^2 + x_2{}^2$$

∴
$$\frac{dV(\mathbf{x})}{dt} = \frac{\partial V}{\partial x_1} x_1 + \frac{\partial V}{\partial x_2} x_2$$

or,
$$\frac{dV(x)}{dt} = 2x_1 x_1 + 2x_2 x_1. \tag{3.11}$$

Putting $\dot{x}_1$ and $\dot{x}_2$ in Equation (3.11), we have

$$\frac{dV(\mathbf{x})}{dt} = 2x_1(x_2) + 2x_2(-x_1 + x_2{}^3)$$

$$= 2x_1 x_2 - 2x_1 x_2 - 2x_2{}^4$$

$$= -2x_2{}^4.$$

Clearly, $\dfrac{dV(\mathbf{x})}{dt} < 0; \mathbf{x} \neq \mathbf{0}$

implies that $\dfrac{dV(\mathbf{x})}{dt}$ is negative-definite.

Therefore, the system under consideration is **asymptotically stable** at the origin. (Refer to Theorem 3.2). **Ans.**

Example 3.3. *Consider a nonlinear system described by*

$$x_1 = x_2$$

$$x_2 = -x_2 - x_1{}^3.$$

Investigate whether the system is stable or not.

Solution. We might choose a positive-definite scalar function as,

$$V(\mathbf{x}) = x_1{}^2 + x_2{}^2$$

∴
$$\frac{dV(\mathbf{x})}{dt} = \frac{\partial V}{\partial x_1} x_1 + \frac{\partial V}{\partial x_2} x_2$$

or,
$$\frac{dV(\mathbf{x})}{dt} = 2x_1 x_1 + 2x_2 x_2. \tag{3.12}$$

Putting $\dot{x}_1$ and $\dot{x}_2$ in Equation (3.12), we have

$$\frac{dV(\mathbf{x})}{dt} = 2x_1(x_2) + 2x_2(-x_2 - x_1^3)$$

or,

$$\frac{dV(\mathbf{x})}{dt} = 2x_1x_2 - 2x_2^2 - 2x_1^3x_2.$$

Here, $\dfrac{dV(\mathbf{x})}{dt}$ is **indefinite** (Refer to Section 3.4.5). So we **cannot predict** the stability of the system. This implies that this particular $V(\mathbf{x})$ is not a suitable Lyapunov function. As we are unable to get the proper Lyapunov function, it no way implies that the system is unstable. It only means that **our attempt in trying to establish the stability** of the system has **failed**. (Refer to Section 3.7).

The choice of a suitable Lyapunov function depends on the **ingenuity** of the designer. Let us choose some other positive-definite scalar function for the same system as

$$V(\mathbf{x}) = x_1^4 + x_1^2 + 2x_1x_2 + 2x_2^2$$

$\therefore$

$$\frac{dV(\mathbf{x})}{dt} = (4x_1^3 + 2x_1 + 2x_2)x_1 + (2x_1 + 4x_2)x_2. \tag{3.13}$$

Putting $\dot{x}_1$ and $\dot{x}_2$ in Equation (3.13), we have

$$\frac{dV(\mathbf{x})}{dt} = (4x_1^3 + 2x_1 + 2x_2)x_2 + (2x_1 + 4x_2)(-x1 - x_2^3)$$

or,

$$\frac{dV(\mathbf{x})}{dt} = -2x_1^4 - 2x_2^2.$$

Clearly, $\dfrac{dV(\mathbf{x})}{dt} < 0; \mathbf{x} \neq 0$

implies that is $\dfrac{dV(\mathbf{x})}{dt}$ negative-definite.

Further, $V(\mathbf{x}) \rightarrow \infty$ as $\|\mathbf{x}\| \rightarrow \infty$.

Therefore, the system under consideration is **asymptotically stable in-the-large** at the origin. (Refer to Theorem 3.3).   **Ans.**

Example 3.4. *Investigate the stability of the system described by*

$$x_1 = x_2$$
$$x_2 = -x_1 - x_1^2 x_2.$$

Solution. We might choose a positive-definite scalar function as

$$V(\mathbf{x}) = x_1^2 + x_2^2$$

$$\therefore \qquad \frac{dV(\mathbf{x})}{dt} = \frac{\partial V}{\partial x_1} x_1 + \frac{\partial V}{\partial x_2} x_2$$

or,
$$\frac{dV(\mathbf{x})}{dt} = 2x_1 x_1 + 2x_2 x_2. \qquad (3.14)$$

Putting $\dot{x}_1$ and $\dot{x}_2$ in Equation (3.14), we have

$$\frac{dV(\mathbf{x})}{dt} = 2x_1 x_2 + 2x_2(-x_1 - x_1^2 x_2)$$

or,
$$\frac{dV(\mathbf{x})}{dt} = 2x_1 x_2 - 2x_1 x_2 - 2x_1^2 x_2^2$$

or,
$$\frac{dV(\mathbf{x})}{dt} = -2x_1^2 x_2^2.$$

Clearly, $\dfrac{dV(\mathbf{x})}{dt} < 0; \mathbf{x} \neq 0$

implies that $\dfrac{dV(\mathbf{x})}{dt}$ is negative-definite.

Further, $V(\mathbf{x}) \to \infty$ as $\|\mathbf{x}\| \to \infty$.

Therefore, the system under consideration is **asymptotically stable in-the-large** at the origin.   **Ans.**

Example 3.5. *Determine whether the system is stable or not, given that*

$$\mathbf{x} = \begin{bmatrix} 0 & 1 \\ -1 & -1 \end{bmatrix} \mathbf{x}.$$

Solution. Given that

$$\begin{bmatrix} x_1 \\ x_2 \end{bmatrix} = \begin{bmatrix} 0 & 1 \\ -1 & -1 \end{bmatrix} \begin{bmatrix} x_1 \\ x_2 \end{bmatrix}.$$

We may write

$$x_1 = x_2$$
$$x_2 = -x_1 - x_2.$$

We might choose a positive-definite scalar function as,

$$V(\mathbf{x}) = x_1^2 + x_2^2$$

$$\therefore \qquad \frac{dV(\mathbf{x})}{dt} = \frac{\partial V}{\partial x_1} x_1 + \frac{\partial V}{\partial x_2} x_2$$

or, $$\qquad \frac{dV(\mathbf{x})}{dt} = 2x_1 x_1 + 2x_2 x_2. \qquad (3.15)$$

Putting $\dot{x}_1$ and $\dot{x}_2$ in Equation (3.15), we have

$$\frac{dV(\mathbf{x})}{dt} = 2x_1 x_2 + 2x_2(-x_1 - x_2)$$

or, $$\qquad \frac{dV(\mathbf{x})}{dt} = -2x_2^2.$$

Clearly, $\dfrac{dV(\mathbf{x})}{dt} < 0; \mathbf{x} \neq 0$

implies that $\dfrac{dV(\mathbf{x})}{dt}$ is negative-definite.

Therefore, the system under consideration is **asymptotically stable** at the origin. **Ans.**

3.8   LYAPUNOV'S DIRECT METHOD AND LINEAR TIME-INVARIANT SYSTEM

Consider a linear autonomous system described by the state Equation

$$\mathbf{x}(t) = \mathbf{A}\mathbf{x}(t) \qquad (3.16)$$

where $\mathbf{x}(t)$ is a state vector and $\mathbf{A}$ is an $n \times n$ constant matrix. We assume that $\mathbf{A}$ is nonsingular, i.e., $|\mathbf{A}| \neq 0$, so that the only equilibrium state is the origin, i.e., $\mathbf{x}_e(t) = \mathbf{0}$. We can easily investigate the stability of the system described by Equation (3.16) by use of the direct method of Lyapunov.

We choose a possible Lyapunov function for the system described by Equation (3.16) as

$$V(\mathbf{x}) = \mathbf{x}^T \mathbf{P} \mathbf{x}$$

where $\mathbf{P}$ is a real, positive-definite, symmetric matrix.

The time-derivative of $V(\mathbf{x})$ along any trajectory is given by

$$\frac{dV(\mathbf{x})}{dt} = \mathbf{x}^T \mathbf{P} \mathbf{x} + \mathbf{x}^T \mathbf{P} \mathbf{x}$$

$$= (\mathbf{A}\mathbf{x})^T \mathbf{P} \mathbf{x} + \mathbf{x}^T \mathbf{P} \mathbf{A} \mathbf{x}$$

$$= \mathbf{x}^T \mathbf{A}^T \mathbf{P} \mathbf{x} + \mathbf{x}^T \mathbf{P} \mathbf{A} \mathbf{x}$$

$$= \mathbf{x}^T (\mathbf{A}^T \mathbf{P} + \mathbf{P} \mathbf{A}) \mathbf{x}.$$

Since $V(\mathbf{x})$ was choosen to be **positive-definite**, for $\dfrac{dV(\mathbf{x})}{dt}$ to be **negative-definite**, we require that

$$\frac{dV(\mathbf{x})}{dt} = -\mathbf{x}^T \mathbf{Q} \mathbf{x}$$

where, $\mathbf{Q} = -(\mathbf{A}^T \mathbf{P} + \mathbf{P} \mathbf{A})$ (3.17)
which is a real, positive-definite, symmetric matrix.

The norm of $\mathbf{x}$ is given by

$$\| \mathbf{x} \| = (\mathbf{x}^T \mathbf{P} \mathbf{x})^{1/2}$$

then, $$V(\mathbf{x}) = \| \mathbf{x} \|^2$$

thus, $$V(\mathbf{x}) \to \infty \text{ as } \| \mathbf{x} \| \to \infty.$$

Hence, the system described by Equation (3.16) is **asymptotically stable in-the-large** at the origin and **for this result to hold**, it is **sufficient** that $\mathbf{Q}$ be **positive-definite**.

Instead of first specifying a positive-definite matrix $\mathbf{P}$ and examining whether $\mathbf{Q}$ is positive-definite, it is more convenient to specify a positive-definite matrix $\mathbf{Q}$ first and then examine whether $\mathbf{P}$, determined from Equation (3.17), is positive-definite. Note that the conditions for the positive-definiteness of **P** are **sufficient** for the system described by (3.16) to be **asymptotically stable in-the-large**. The conditions are **necessary** also and in order to show this fact, suppose that the system described by (3.16) is asymptotically stable in-the-large at the origin and **P** is negative-definite.

Consider the scalar function

$$V(\mathbf{x}) = \mathbf{x}^T \mathbf{P} \mathbf{x}$$

therefore,
$$\frac{dV(\mathbf{x})}{dt} = -(\mathbf{x}^T \mathbf{P} \mathbf{x} + \mathbf{x}^T \mathbf{P} \mathbf{x}) \tag{3.18}$$

$$= -[(\mathbf{A}\mathbf{x})^T \mathbf{P} \mathbf{x} + \mathbf{x}^T \mathbf{P} \mathbf{A} \mathbf{x}]$$

$$= -[\mathbf{x}^T \mathbf{A}^T \mathbf{P} \mathbf{x} + \mathbf{x}^T \mathbf{P} \mathbf{A} \mathbf{x}]$$

$$= -[\mathbf{x}^T (\mathbf{A}^T \mathbf{P} + \mathbf{P} \mathbf{A}) \mathbf{x}]$$

$$= \mathbf{x}^T \mathbf{Q} \mathbf{x}$$

$$> 0.$$

This shows a **contradiction** since $V(\mathbf{x})$ given by (3.18) satisfies the **instability theorem** (Refer to Theorem 3.4).

Therefore, we can conclude that the conditions for the positive-definiteness of **P** are **necessary and sufficient** for the system described by (3.16) to be **asymptotically stable in-the-large** at the origin. We shall summarize what we have just stated in the form of a theorem given next.

Theorem 3.5. Consider a linear autonomous system, described by the state equation

$$\mathbf{x}(t) = \mathbf{A}\mathbf{x}(t) \tag{3.19}$$

where $\mathbf{x}(t)$ is a state-vector and **A** is an $n \times n$ constant nonsingular matrix.

The linear system described by state Equation (3.19) be **asymptotically stable in-the-large** at the origin (equilibrium point) **if, and only if,** given any real, positive-definite, symmetric matrix **Q** there exists a real, positive-definite, symmetric matrix **P** such that

$$\mathbf{A}^T \mathbf{P} + \mathbf{P} \mathbf{A} = -\mathbf{Q}.$$

The scalar function $\mathbf{x}^T\mathbf{P}\mathbf{x}$ being a **Lyapunov function** for the system described by Equation (3.19).

Example 3.6. *Consider the system described by*

$$\mathbf{x} = \mathbf{A}\mathbf{x}$$

$$\mathbf{A} = \begin{bmatrix} 0 & 1 \\ -1 & -1 \end{bmatrix}.$$

Determine whether the system is stable or not.

Solution. Let us assume a tentative Lyiapunov function as

$$V(\mathbf{x}) = \mathbf{x}^T\mathbf{P}\mathbf{x}$$

where **P** is to be obtained by solving the equation

$$\mathbf{A}^T\mathbf{P} + \mathbf{P}\mathbf{A} = -\mathbf{Q} \qquad (3.20)$$

for an arbitrary choice of positive-definite, real-symmetric matrix **Q**. It is convenient to choose **Q** = **I**, the **identity matrix**. Equation (3.20) then become

$$\mathbf{A}^T\mathbf{P} + \mathbf{P}\mathbf{A} = -\mathbf{I}$$

or,
$$\begin{bmatrix} 0 & -1 \\ 1 & -1 \end{bmatrix}\begin{bmatrix} p_{11} & p_{12} \\ p_{21} & p_{22} \end{bmatrix} + \begin{bmatrix} p_{11} & p_{12} \\ p_{21} & p_{22} \end{bmatrix}\begin{bmatrix} 0 & 1 \\ -1 & -1 \end{bmatrix} = \begin{bmatrix} -1 & 0 \\ 0 & -1 \end{bmatrix}.$$

As matrix **P** is known to be a positive-definite, real-symmetric matrix for a stable system, we may take $p_{12} = p_{21}$.

Therefore,

$$\begin{bmatrix} 0 & -1 \\ 1 & -1 \end{bmatrix}\begin{bmatrix} p_{11} & p_{12} \\ p_{21} & p_{22} \end{bmatrix} + \begin{bmatrix} p_{11} & p_{12} \\ p_{21} & p_{22} \end{bmatrix}\begin{bmatrix} 0 & 1 \\ -1 & -1 \end{bmatrix} = \begin{bmatrix} -1 & 0 \\ 0 & -1 \end{bmatrix}.$$

By expanding this matrix equation, we obtain three simultaneous equations as follows

$$-2p_{12} = -1,$$
$$p_{11} - p_{12} - p_{22} = 0,$$
$$2p_{12} - 2p_{22} = -1.$$

Solving for p_{11}, p_{12}, p_{22} we obtain

$$\mathbf{P} = \begin{bmatrix} p_{11} & p_{12} \\ p_{12} & p_{22} \end{bmatrix} = \begin{bmatrix} 3/2 & 1/2 \\ 1/2 & 1 \end{bmatrix}.$$

Now, applying Sylvestor's theorem, in order to test the positive-definiteness of $\mathbf{P}$. We have

$$p_{11} = 3/2 > 0$$

$$\begin{vmatrix} p_{11} & p_{12} \\ p_{12} & p_{22} \end{vmatrix} = \begin{vmatrix} 3/2 & 1/2 \\ 1/2 & 1 \end{vmatrix} = \frac{5}{4} > 0.$$

Since all the successive principal minors of the matrix $\mathbf{P}$ are positive, this means that $\mathbf{P}$ is positive-definite. Hence, the system under consideration is **asymptotically stable in-the-large** at the origin and the Lyapunov function is in this case

$$V(\mathbf{x}) = \mathbf{x}^T \mathbf{P} \mathbf{x}$$

or,

$$V(\mathbf{x}) = \begin{bmatrix} x_1 & x_2 \end{bmatrix} \begin{bmatrix} 3/2 & 1/2 \\ 1/2 & 1 \end{bmatrix} \begin{bmatrix} x_1 \\ x_2 \end{bmatrix}$$

or,

$$V(\mathbf{x}) = \frac{1}{2}(3x_1^2 + 2x_1x_2 + 2x_2^2)$$

and,

$$\frac{dV(\mathbf{x})}{dt} = \frac{\partial V}{\partial x_1} x_1 + \frac{\partial V}{\partial x_2} x_2$$

$$= \frac{1}{2}(6x_1 + 2x_2)x_1 + \frac{1}{2}(2x_1 + 4x_2)x_2$$

$$= (3x_1 + x_2)(x_2) + (x_1 + 2x_2)(-x_1 - x_2)$$

$$= -(x_1^2 + x_2^2) \qquad \textbf{Ans.}$$

3.9   CONSTRUCTING LYAPUNOV'S FUNCTION FOR NONLINEAR SYSTEMS (KRASOVSKII'S METHOD)

As we have seen in earlier sections, the Lyapunov theorems give only sufficient conditions on system stability and furthermore, there is no unique way of constructing a Lyapunov function except in the case of linear systems (refer to Section 3.8), where a Lyapunov function can always be constructed and both necessary and sufficient conditions established. We shall now present **Krasovskii's method** of constructing a Lyapunov function for nonlinear systems. Consider the system

$$\mathbf{x}(t) = \mathbf{F}(\mathbf{x}(t)); \mathbf{F}(0) = 0. \tag{3.20}$$

Assume that $\mathbf{F}(\mathbf{x})$ is differentiable with respect to all components of $\mathbf{x}$. Then, the **Jacobian matrix** for the system described by (3.21) is given by

$$\mathbf{J}(\mathbf{x}) = \begin{bmatrix} \dfrac{\partial F_1}{\partial x_1} & \dfrac{\partial F_1}{\partial x_2} & \dfrac{\partial F_1}{\partial x_n} \\[2mm] \dfrac{\partial F_2}{\partial x_1} & \dfrac{\partial F_2}{\partial x_2} & \dfrac{\partial F_2}{\partial x_n} \\[4mm] \dfrac{\partial F_n}{\partial x_1} & \dfrac{\partial F_n}{\partial x_2} & \dfrac{\partial F_n}{\partial x_n} \end{bmatrix}_{n \times n}.$$

Define also

$$\mathbf{K}(\mathbf{x}) = \mathbf{J}^T(\mathbf{x}) + \mathbf{J}(\mathbf{x}). \tag{3.22}$$

If $\mathbf{K}(\mathbf{x})$ is negative-definite, then

$$|\mathbf{K}(\mathbf{x})| \neq 0 \text{ for } \mathbf{x} \neq 0.$$

Thus, there is no other equilibrium state than $\mathbf{x} = \mathbf{0}$ in the entire state-space, i.e.,

$$\mathbf{F}(\mathbf{x}) \neq 0 \text{ for } \mathbf{x} \neq 0 \tag{3.23}$$

If we observe (3.21) and (3.23), we find that a scalar function $V(\mathbf{x})$ given by

$$V(\mathbf{x}) = \mathbf{F}^T(\mathbf{x})\mathbf{F}(\mathbf{x}) \tag{3.24}$$

is positive-definite.

Now,

$$\mathbf{F}(\mathbf{x}) = \frac{\partial \mathbf{F}(\mathbf{x})}{\partial t}$$

or,

$$\mathbf{F}(\mathbf{x}) = \frac{\partial \mathbf{F}(\mathbf{x})}{\partial \mathbf{x}(t)} \cdot \frac{\partial \mathbf{x}(t)}{\partial t}$$

or,

$$\mathbf{F}(\mathbf{x}) = \mathbf{J}(\mathbf{x})\,\mathbf{F}(\mathbf{x}). \tag{3.25}$$

We can obtain $\dot{V}(\mathbf{x})$ as

$$
\begin{aligned}
V(\mathbf{x}) &= \mathbf{F}^T(\mathbf{x})\,\mathbf{F}(\mathbf{x}) + \mathbf{F}^T(\mathbf{x})\,\mathbf{F}(\mathbf{x}) \\
&= [\mathbf{J}(\mathbf{x})\,\mathbf{F}(\mathbf{x})]^T\,\mathbf{F}(\mathbf{x}) + \mathbf{F}^T(\mathbf{x})\,\mathbf{J}(\mathbf{x})\,\mathbf{F}(\mathbf{x}) \\
&= \mathbf{F}^T(\mathbf{x})[\mathbf{J}^T(\mathbf{x}) + \mathbf{J}(\mathbf{x})]\,\mathbf{F}(\mathbf{x}) \\
&= \mathbf{F}^T(\mathbf{x})\,\mathbf{K}(\mathbf{x})\,\mathbf{F}(\mathbf{x}).
\end{aligned}
$$

Obviously, if $\mathbf{K}(\mathbf{x})$ is negative-definite, it follows that $\dot{V}(\mathbf{x})$ is negative-definite.

Hence, $V(\mathbf{x})$ is a **Lyapunov function** and therefore, the system described by (3.21) is **asymptotically stable** at the origin. Moreover, if $V(\mathbf{x}) = \mathbf{F}^T(\mathbf{x})\mathbf{F}(\mathbf{x})$ tends to infinity as $\|\mathbf{x}\| \to \infty$, then the system described by (3.21) is **asymptotically stable in-the-large** at the origin.

Now, we shall summarize what we have just stated in the form of a theorem, called **Krasovskii's Theorem**, given next.

Theorem 3.6. (Krasovskii's Theorem):

Consider the system described by

$$\mathbf{x}(t) = \mathbf{F}(\mathbf{x}(t));\ \mathbf{F}(0) = 0. \tag{3.26}$$

Assume that $\mathbf{F}(\mathbf{x})$ is differentiable with respect to all components of $\mathbf{x}$. The Jacobian matrix for the system is

$$
\mathbf{J}(\mathbf{x}) =
\begin{bmatrix}
\dfrac{\partial F_1}{\partial x_1} & \dfrac{\partial F_1}{\partial x_2} & & \dfrac{\partial F_1}{\partial x_n} \\[2ex]
\dfrac{\partial F_2}{\partial x_1} & \dfrac{\partial F_2}{\partial x_2} & & \dfrac{\partial F_2}{\partial x_n} \\[2ex]
& & & \\[1ex]
\dfrac{\partial F_n}{\partial x_1} & \dfrac{\partial F_n}{\partial x_2} & & \dfrac{\partial F_n}{\partial x_n}
\end{bmatrix}_{n \times n}.
$$

Define also

$$\mathbf{K}(\mathbf{x}) = \mathbf{J}^T(\mathbf{x}) + \mathbf{J}(\mathbf{x}).$$

If $\mathbf{K}(\mathbf{x})$ is negative-definite, then the system described by (3.26) is **asymptotically stable** at the origin. A Lyapunov function for this system is

$$V(\mathbf{x}) = \mathbf{F}^T(\mathbf{x})\mathbf{F}(\mathbf{x}).$$

If, in addition,

$$\mathbf{F}^T(\mathbf{x})\mathbf{F}(\mathbf{x}) \to \infty \text{ as } \| \mathbf{x} \| \to \infty,$$

then the system, described by (3.26) is **asymptotically stable in-the-large** at the origin.

Krasovskii's theorem gives a **sufficient condition** of asymptotic stability for nonlinear systems. Notice that Krasovskii's theorem differs from the usual linearization approaches. It is not limited to small departures from the equilibrium state. Moreover, failure to obtain a suitable Lyapunov function by this method **does not imply** instability of the system.

Example 3.7. *By use of Krasovskii's theorem, determine whether the system is stable or not, given*

$$x_1 = -x_1$$
$$x_2 = x_1 - x_2 - x_2^3.$$

Also determine the Lyapunov function for this system.

Solution. Given,

$$\mathbf{x}(t) = \begin{bmatrix} -x_1 \\ x_1 - x_2 - x_2^3 \end{bmatrix} = \mathbf{F}(\mathbf{x}).$$

Now, $$\mathbf{J}(\mathbf{x}) = \begin{bmatrix} \dfrac{\partial F_1}{\partial x_1} & \dfrac{\partial F_1}{\partial x_2} \\ \dfrac{\partial F_2}{\partial x_1} & \dfrac{\partial F_2}{\partial x_2} \end{bmatrix} = \begin{bmatrix} -1 & 0 \\ 1 & (-1 - 3x_2^2) \end{bmatrix}.$$

Hence, $\mathbf{K}(\mathbf{x}) = \mathbf{J}^T(\mathbf{x}) + \mathbf{J}(\mathbf{x})$

$$= \begin{bmatrix} -1 & 1 \\ 0 & (-1-3x_2^{\,2}) \end{bmatrix} + \begin{bmatrix} -1 & 0 \\ 1 & (-1-3x_2^{\,2}) \end{bmatrix}$$

$$= \begin{bmatrix} -2 & 1 \\ 1 & (-2-6x_2^{\,2}) \end{bmatrix}.$$

Hence, $\mathbf{K}(\mathbf{x})$ is negative-definite for all $\mathbf{x} \neq \mathbf{0}$. The Lyapunov function for this system is

$$V(\mathbf{x}) = \mathbf{F}^T(\mathbf{x})\,\mathbf{F}(\mathbf{x})$$

$$= [-x_1 \quad (x_1 - x_2 - x_2^{\,3})] \begin{bmatrix} -x_1 \\ x_1 - x_2 - x_2^{\,3} \end{bmatrix} \qquad \textbf{Ans.}$$

$$= x_1 + (x_1 - x_2 - x_2^{\,3})^2$$

Furthermore, $V(\mathbf{x}) \to \infty$ as $\| \mathbf{x} \| \to \infty$.

We conclude that, the system under consideration is **asymptotically stable in-the-large** at the origin. **Ans.**

Readers should note that $\mathbf{K}(\mathbf{x})$ being negative-definite requires that $\mathbf{J}(\mathbf{x})$ must have nonzero elements on its main diagonal.

3.10 POPOV'S CRITERION FOR STABILITY OF NONLINEAR SYSTEMS

V.M. Popov obtained a **frequency domain stability criterion**, which is quite similar to the Nyquist criterion, as a sufficient condition for **asymptotic stability** of an important class of nonlinear systems. Popov's stability criterion applies to a closed-loop control system that consists of a nonlinear element and a linear time-invariant plant, as shown in Figure 3.2. The **nonlinearity**

is described by a functional relation that must lie in the first and the third quadrants, as shown in Figure 3.3.

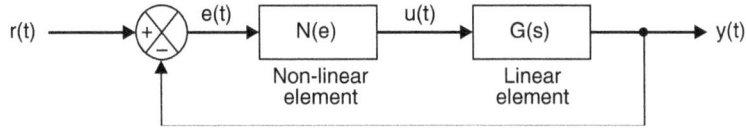

FIGURE 3.2   Popov's basic feedback control system.

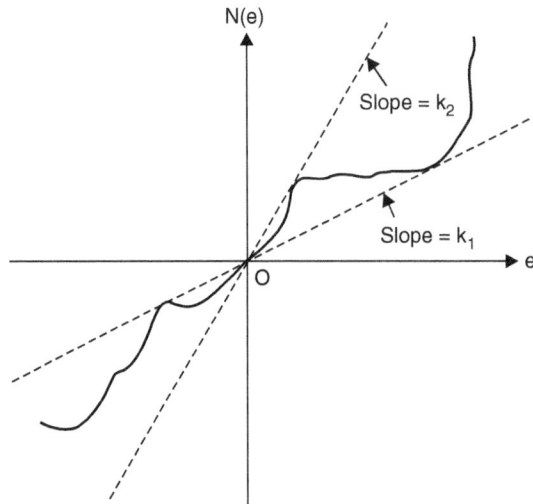

FIGURE 3.3   Nonlinear characteristics.

Many control systems with a single nonlinearity, in practice, can be modeled by the block-diagram and the nonlinear characteristics of Figures 3.2 and 3.3, respectively.

Popov's stability criterion is based on the following **assumptions**:

1. The transfer function $G(s)$ of the linear part of the system, has **more poles than zeros** and there are **no pole-zero cancellations**.

2. The nonlinear characteristic is **bounded** by k_1 and k_2 as shown in Figure 3.3, i.e.,

$$k_1 \leq N(e) \leq k_2. \tag{3.27}$$

Now, the **Popov's Criterion** is stated as follows:

*"The closed-loop system is **asymptotically stable** if the Nyquist plot of G(jω) does not intersect or enclose the circle, which is described by*

$$\left[x + \frac{k_1 + k_2}{2k_1 k_2} \right]^2 + y^2 = \left[\frac{k_2 - k_1}{2k_1 k_2} \right]^2. \tag{3.28}$$

where x and y denote the real and imaginary coordinates of the G(jω)-plane, respectively."

Note that the Popov's Stability Criterion is **sufficient but is not necessary**. If the preceeding condition is violated, it does not necessarily mean that the system is unstable. Popov's stability criterion can be illustrated gemetrically, as shown in Figure 3.4.

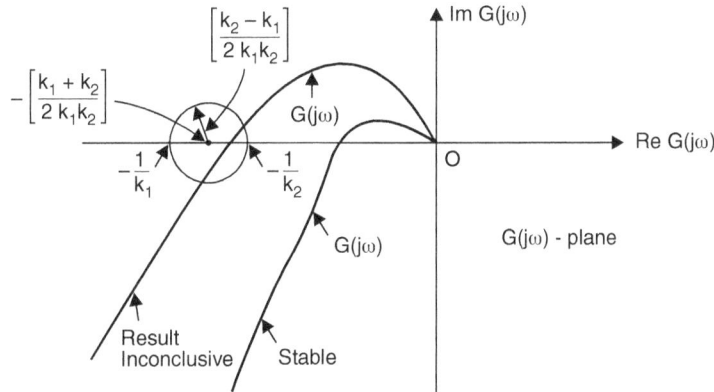

FIGURE 3.4   Geometrical interpretation of Popov's criterion.

In practice, the system nonlinearities (in a great majority) are with $k_1 = 0$. In that case, the circle of Equation (3.28) is replaced by a straight line, given by

$$x = -\frac{1}{k_2}. \tag{3.29}$$

For stability, the Nyquist plot of $G(j\omega)$ must not intersect this line. This fact is illustrated in Figure 3.5.

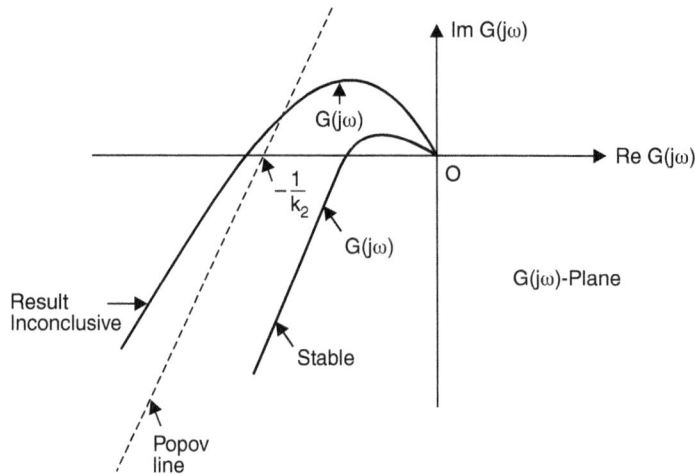

FIGURE 3.5   Popov's criterion for common nonlinearities.

For linear systems, $k_1 = k_2 = k$ and it is interesting to observe that the circle of Equation (3.28) **degenerates** to the $(-1, j0)$ point and Popov's stability criterion becomes **Nyquist's Criterion**.

EXERCISES

1. Consider a nonlinear system described by:

$$\dot{x}_1 = -x_1 + 2x_1^2 x_2$$
$$\dot{x}_2 = -x_2.$$

Investigate whether the system is stable or not.

2. Consider a second-order system with two nonlinearties:

$$\dot{x}_1 = f_1(x_1) + f_2(x_2)$$
$$\dot{x}_2 = x_1 + ax_2.$$

Assume that $f_1(0) = f_2(0) = 0$ and that $f_1(x_1)$ and $f_2(x_2)$ are real and differentiable. In addition,

$$[f_1(x_1) + f_2(x_2)]^2 + [x_1 + ax_2]^2 \to \infty, \text{ as } \|\mathbf{x}\| \to \infty.$$

Determine sufficient conditions for asymptotic stability of the equilibrium state $x = 0$.

[**Hint.** Use Krasovskii's theorem.]

3. A linear system is descibed by

$$x = Ax$$

where, $A = \begin{bmatrix} -1 & -2 \\ 1 & -4 \end{bmatrix}$.

Determine whether the system is stable or not.

4. A nonlinear system is described by:

$$x_1 = -3x_1 + x_2$$
$$x_2 = x_1 - x_2 - x_2^{3}.$$

Investigate the stability of the equilibrium state.

5. Consider the system described by:

$$x = \begin{bmatrix} 0 & 1 \\ -1 & -2 \end{bmatrix}.$$

Determine the stability of the equilibrium state

$$x = 0.$$

6. A nonlinear system is governed by the equations

$$x_1 = x_2$$
$$x_2 = -x_1 - (1 - |x_1|) x_2.$$

Determine the region in the state-plane for which the equilibrium state $x = 0$ is asymptotically stable.

REFERENCES

(1) A.W. Langill Jr., *Automatic Control Systems Engineering*, Prentice-Hall, Inc., Englewood Cliffs, New Jersey, 2nd Volume, 1965.

(2) D. Roy Choudhury, *Modern Control Engineering*, Prentice-Hall of India Pvt. Ltd, New Delhi, 2005.

(3) I.J. Nagrath, M. Gopal, *Control Systems Engineering*, New Age International Publishers, New Delhi, 4th Edition, 2005.

(4) Katsuhiko Ogata, *Modern Control Engineering*, Prentice-Hall, Upper Saddle River, New Jersey, 3rd Edition, 1997.

4

OPTIMAL CONTROL

4.1 INTRODUCTION

The design of a control system is an attempt to meet a set of specifications that define the overall performance of the system in terms of certain measurable quantities. In the classical design method of control systems, the designer is given a set of specifications in time domain or in frequency domain with the system configuration. Peak overshoot, settling time, gain-margin, phase-margin, steady-state error, etc., are among the most commonly used specifications. These specifications have to be satisfied simultaneously in design. In practice, it may not be possible to satisfy all the desired specifications and hence, the design necessarily becomes a **trial – and – error procedure**. This trial – and – error design procedure works satisfactorily for single-input-single-output systems. However, for a system with multi-input-multi-output having a high degree of complexity, the trial – and – error approach may not lead to a satisfactory design.

The optimal control design is aimed at obtaining a **best possible system** of a particular type with respect to a certain **performance index** or **design criterion**—hence the word 'optimal.' In the optimal control design, the performance index replaces the conventional design criteria, such as peak overshoot, settling time, gain-margin, phase-margin, steady-state error, etc. Of course, the designer must be able to select the performance index properly so that one may describe the **goodness** of the system response on the basis of this performance index.

4.2   PERFORMANCE INDICES

A performance index is a **function** of the **variable system parameters**. It is a single **measure** of a system's performance that emphasises those characteristics of the response that are deemed to be important.

The idea of a performance index is very important in optimal control theory where **extremum** (minimum or maximum) value of this index corresponds to the optimum set of parameter values of the system to be designed. Other desirable features of a performance index are its **selectivity**, i.e., its ability to clearly distinguish between an optimum and nonoptimum system, its **sensitivity** to parameter variations, and the ease of its analytical computation or digital determination.

A number of such performance indices are used in practice, the most common being ISE, i.e., the **integral square error**, given by

$$\text{ISE} = \int_0^\infty e^2(t)\, dt. \tag{4.1}$$

Apart from the ease of implementation, this index has the advantage that it is mathematically convenient both for analysis and computation. By focusing on the **square of the error function**, it **penalizes** the positive as well as negative values of the error.

Another useful performance index is IAE, i.e., **integral of the absolute magnitude of error** given by

$$\text{IAE} = \int_0^\infty |e(t)|\, dt. \tag{4.2}$$

By focusing on the **magnitude of the error**, it **penalizes** either the positive or negative values of the error.

In order to penalize the long duration transients, the following indices are proposed:

1. **Integral time-absolute error**, given by

$$\text{ITAE} = \int_0^\infty t\,|e(t)|\, dt \tag{4.3}$$

2. **Integral time-square error**, given by

$$\text{ITSE} = \int_0^\infty t e^2(t)\, dt \tag{4.4}$$

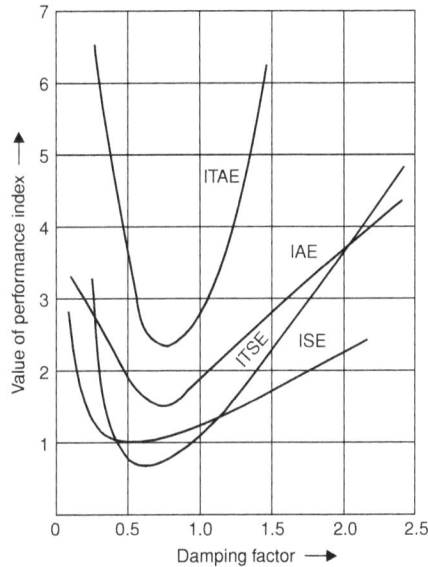

FIGURE 4.1   **Comparison of performance-indices for a second-order system of Equation (4.5).**

For a second-order system having the transfer function

$$G(s) = \frac{1}{s^2 + 2s + 1},$$

(4.5)

the comparison of different performance indices is shown in Figure 4.1.

On observation, we find that the ISE curve is rather flat near the point where the performance index is minimum, i.e., the point corresponding to the damping factor of 0.5. Therefore, the selectivity of ISE is poor. The IAE performance index, which has a minimum for a damping factor of 0.7, gives **slightly better selectivity** than ISE. The ITAE performance index, which has a minimum for a damping factor of 0.707, produces smaller overshoots and oscillations than the IAE and ISE performance indices. In addition, it is the **most sensitive** of the three, i.e., it has the **best selectivity**. The ITSE performance index is **somewhat less sensitive** but is not comfortable computationally.

In practice, relatively **insensitive criterion** may be more useful and hence, based on this logic, ISE may be the most desirable performance index. However, if selectivity is more important, the ITAE performance index is an **obvious choice**.

There are some other performance indices, such as the **integral of squared time-square error** (ISTSE) and the **integral of squared time-absolute error** (ISTAE). These performance indices are given by

$$\text{ISTSE} = \int_0^\infty t^2 e^2(t)\, dt, \tag{4.6}$$

$$\text{ISTAE} = \int_0^\infty t^2 |e(t)|\, dt. \tag{4.7}$$

However, these performance indices are not applied in practice due to the increased difficulty in handing them.

4.3   OPTIMAL CONTROL PROBLEMS

As stated earlier, an optimal control problem is basically associated with the design of a **best possible control system of a particular type** with respect to a certain performance index or design criterion. A control system is optimum, when the selected performance index is **minimized**. We have two approaches for optimal control design problems, namely

1. Transfer function approach, and
2. State-variable approach.

The transfer function approach is straightforward for the optimal control problems in single-input-single-output linear time-invariant systems. However, the limitation of this approach appears when dealing with multi-input-multi-output linear time-invariant systems. In addition, the transfer function design approach becomes ineffective for time-varying and nonlinear systems. Furthermore, the transfer function approach is restricted to systems with a **quadratic performance index** with **no constraint** on time.

All these limitations of the transfer the function approach are absent in the state-variable approach. With the availability of digital computers, the state-variable approach has become the most powerful method for solving modern complex control problems. We shall formulate various optimal control problems in the following sections using the state-variable approach only.

The following steps are involved in the solution of an optimal control problem using the state-variable approach:

(*a*) Given a plant, described by

$$\mathbf{x}(t) = \mathbf{A}\mathbf{x}(t) + \mathbf{B}\mathbf{u}(t) \tag{4.8 A}$$

$$\mathbf{y}(t) = \mathbf{C}\mathbf{x}(t). \tag{4.8 B}$$

find the control law $\mathbf{u}^*(t)$, which is optimal with respect to a given performance index.

(b) Realize the control law that is obtained from step (a).

Let us now discuss some typical control problems in order to provide some physical motivation for the selection of a performance index.

4.3.1   The Finite-Time State Regulator Problem

The **design objective** is to keep the system states at the **equilibrium state** and the system should be able to return to the equilibrium state from any initial state.

Without loss of generality, we assume the origin of state-space as the equilibrium state. Therefore, if the state $\mathbf{x}(t)$ of a system described by Equations (4.8) is required to be close to the equilibrium state $\mathbf{x}_e(t) = 0$, then a reasonable performance index is the **integral-square error**, given by

$$J_e = \int_{t_o}^{t_f} [\mathbf{x}(t) - \mathbf{x}_e(t)]^T \, \mathbf{Q}[\mathbf{x}(t) \, \mathbf{x}_e(t)]dt$$

or,
$$J_e = \int_{t_o}^{t_f} \mathbf{x}^T(t) \, \mathbf{Q}\mathbf{x}(t)dt \qquad (4.9)$$

where, $\mathbf{Q}$ is a real, symmetric, positive-definite (semidefinite) constant matrix and represents the amount of weight the designer places on the **constraint** on state-variable $\mathbf{x}(t)$.

In order to minimize the deviation of the final state $\mathbf{x}(t_f)$ of the system from the equilibrium state $\mathbf{x}_e(t) = \mathbf{0}$, a possible performance index is

$$J_d = \mathbf{x}^T(t_f) \, \mathbf{H}\mathbf{x}(t_f), \qquad (4.10)$$

where $\mathbf{H}$ is a real, symmetric, positive-definite (semidefinite), constant matrix.

Now, the optimal design obtained by minimizing

$$J = \frac{1}{2}(J_d + J_e)$$

or,
$$J = \frac{1}{2}\mathbf{x}^T(t_f) \, \mathbf{H}\mathbf{x}(t_f) + \frac{1}{2}\int_{t_o}^{t_f} \mathbf{x}^T(t) \, \mathbf{Q}\mathbf{x}(t)dt \qquad (4.11)$$

may be **unsatisfactory** in practice. We can modify the performance index (4.11) by adding a **penalty term** for **physical constraints**, given by

$$J_u = \int_{t_o}^{t_f} \mathbf{u}^T(t)\, \mathbf{R} \mathbf{u}(t) dt \qquad (4.12)$$

where $\mathbf{R}$ is a real, symmetric, positive-definite (or semidefinite), constant matrix.

For the finite-time state-regulator problem, a useful performance index is therefore,

$$J = \frac{1}{2}(J_d + J_e + J_u)$$

or, $$J = \frac{1}{2}\mathbf{x}^T(t_f)\, \mathbf{H}\mathbf{x}(t_f) + \frac{1}{2}\int_{t_o}^{t_f} [\mathbf{x}^T(t)\, \mathbf{Q}\mathbf{x}(t) + \mathbf{u}^T(t)\, \mathbf{R}\mathbf{u}(t)]dt. \qquad (4.13)$$

Note that multiplication by $\frac{1}{2}$ in the preceeding equation does not affect the minimization problem, rather it helps us in mathematical manipulations.

We shall now formulate the finite-time state-regulator problem, as follows : Given a plant, described by

$$\mathbf{x}(t) = \mathbf{A}\mathbf{x}(t) + \mathbf{B}\mathbf{u}(t) \qquad (4.14\ A)$$
$$\mathbf{y}(t) = \mathbf{C}\mathbf{x}(t). \qquad (4.14\ B)$$

find optimal control law u*(t); $t_o \le t \le t_f$, where t_o and t_f are specified initial and final times respectively, so that the quadratic performance index,

$$J = \frac{1}{2}\mathbf{x}^T(t_f)\mathbf{H}\mathbf{x}(t_f) + \frac{1}{2}\int_{t_o}^{t_f} [\mathbf{x}^T(t)\, \mathbf{Q}\mathbf{x}(t) + \mathbf{u}^T(t)\, \mathbf{R}\mathbf{u}(t)]\, dt \qquad (4.15)$$

is minimized, subject to the initial condition,

$$\mathbf{x}(t_0) = \mathbf{x}_0.$$

4.3.2   The Infinite-Time State-Regulator Problem

If the terminal time t_f is **not constrained**, i.e., $t_f \to \infty$, then the system state should approach the equilibrium state $\mathbf{x}_e(t) = \mathbf{0}$ (assuming a stable system), i.e., when $t_f \to \infty$, $\mathbf{x}(\infty) \to \mathbf{0}$ for the optimal system to be stable. Therefore, the terminal constraint has **no significance** in the performance index given by (4.15). Thus, setting $\mathbf{H} = \mathbf{0}$ in the quadratic performance index (4.15), we get the performance index for the infinite-time state-regulator problem as

$$J = \frac{1}{2} \int_{t_0}^{\infty} [\mathbf{x}^T(t)\,\mathbf{Q}\mathbf{x}(t) + \mathbf{u}^T(t)\,\mathbf{R}\mathbf{u}(t)]\,dt. \qquad (4.16)$$

We will now formulate the infinite-time state-regulator problem as follows: Given a plant, described by

$$\mathbf{x}(t) = \mathbf{A}\mathbf{x}(t) + \mathbf{B}\mathbf{u}(t) \qquad (4.17\ A)$$

$$\mathbf{y}(t) = \mathbf{C}\mathbf{x}(t), \qquad (4.17\ B)$$

find optimal control law u*(t); $t_0 \le t < \infty$, where t_0 is specified initial time, so that the quadratic performance index,

$$J = \frac{1}{2} \int_{t_0}^{\infty} [\mathbf{x}^T(t)\,\mathbf{Q}\mathbf{x}(t) + \mathbf{u}^T(t)\,\mathbf{R}\mathbf{u}(t)]\,dt \qquad (4.18)$$

is minimized, subject to the initial condition,

$$\mathbf{x}(t_0) = \mathbf{x}_0.$$

In the finite-time state-regulator problem, there is no restriction on **controllability** of the plant because J is always finite and instability does not impose any problem in finite-interval control. But in the case of infinite-interval control, J can become **infinite** if

(a) one or more states of the plant are **not controllable**,
(b) the uncontrollable states are **unstable**, and
(c) the unstable states are **reflected** in the system's performance index J.

If the performance index J is infinite for all controls, then we cannot distinguish the optimal control from the other controls. Therefore, for the infinite-time-state-regulator problem, **the solution exists only if** the plant described by Equations (4.17) is **completely controllable**, i.e., the performance index J is **finite**.

4.3.3   The Output-Regulator Problem

The **design objective** is to keep the plant output at the **steady-state output**. The output-regulator problem, thus, can be formulated as follows :

Given a plant, described by

$$\mathbf{x}(t) = \mathbf{A}\mathbf{x}(t) + \mathbf{B}\mathbf{u}(t) \qquad (4.19\ A)$$

$$\mathbf{y}(t) = \mathbf{C}\mathbf{x}(t). \qquad (4.19\ B)$$

find optimal control law $\mathbf{u}^*(t)$; $t_o \le t \le t_f$, where t_o and t_f are specified initial and final times respectively, so that the quadratic performance index,

$$J = \frac{1}{2}\mathbf{y}^T(t_f)\,\mathbf{H}\mathbf{y}(t_f) + \frac{1}{2}\int_{t_o}^{t_f}[\mathbf{y}^T(t)\mathbf{Q}\mathbf{y}(t) + \mathbf{u}^T(t)\mathbf{R}\mathbf{u}(t)]\,dt \qquad (4.20)$$

is minimized, subject to the initial condition,

$$\mathbf{x}(t_0) = \mathbf{x}_0.$$

If the plant described by Equations (4.19) is **observable**, then we can **reduce** the output regulator problem to the **state-regulator problem** in order to find the solution to this output-regulator problem.

4.3.4   The Tracking Problem

The **design objective** is to maintain the plant output **as close as possible to the desired output**. We can formulate the traking problem as follows:

Given a plant, described by

$$\mathbf{x}(t) = \mathbf{A}\mathbf{x}(t) + \mathbf{B}\mathbf{u}(t) \qquad (4.\ 21\ A)$$

$$\mathbf{y}(t) = \mathbf{C}\mathbf{x}(t), \qquad (4.21\ B)$$

if $\mathbf{r}(t)$ is the **desired output,** then the tracking error $\mathbf{e}(t)$ is given by

$$\mathbf{e}(t) = \mathbf{y}(t) - \mathbf{r}(t). \qquad (4.22)$$

Find optimal control law $\mathbf{u}^*(t)$, $t_o \le t \le t_f$, where t_o and t_f are specified initial and final times respectively, so that the quadratic performance index,

$$J = \frac{1}{2}\mathbf{e}^T(t_f)\,\mathbf{H}\mathbf{e}(t_f) + \frac{1}{2}\int_{t_o}^{t_f}[\mathbf{e}^T(t)\,\mathbf{Q}\mathbf{e}(t) + \mathbf{u}^T(t)\,\mathbf{R}\mathbf{u}(t)]\,dt \qquad (4.23)$$

is minimized.

4.4   MINIMIZATION OF PERFORMANCE INDEX

As already stated, a **control system is optimum** when the selected **performance index is minimized**. Once the performance index for a system has been choosen, the next task is to determine a **control law** that minimizes this index. With the background material we have accumulated, we can state **explicitely** an optimal control problem as :

Given a plant, described by

$$\mathbf{x}(t) = f(\mathbf{x}(t), \mathbf{u}(t), t), \qquad (4.24)$$

find optimal control law $\mathbf{u}^*(t)$, so that the quadratic performance index,

$$J = \psi(\mathbf{x}(t_f), t_f) + \int_t^{t_f} L(\mathbf{x}(\tau), \mathbf{u}(\tau), \tau) \, d\tau \qquad (4.25)$$

$(t \leq \tau \leq t_f)$; is minimized, subject to the initial condition $\mathbf{x}(t)$.

We have two approaches for accomplishing this minimization, namely

1. **Minimum principle of Pontryagin** (based on the concepts of **calculus of variations**), and
2. **Dynamic programming** developed by **Bellman** (making use of the **Principle of Optimality**).

However, we shall use the **dynamic programming approach** in our discussion.

4.5   PRINCIPLE OF OPTIMALITY

It states that an optimal control strategy has the **property** that whatever the initial state and the control law of the initial stages, the remaining control **must form an optimal** control with respect to the state resulting from the control of the initial stages.

We can explain the principle of optimality with the help of Figure 4.2. If we have found the **optimal trajectory** on the interval from $[t_o, t_f]$ by solving the optimal control problem on that interval, the resulting trajectory is **also optimal on all subintervals** of this interval of the form $[t, t_f]$ with $t > t_o$, provided that the initial condition at time t was obtained from running the system forward along the optimal trajectory from time t_o.

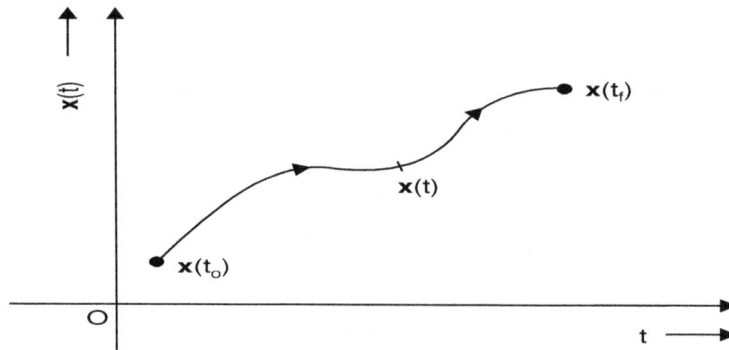

FIGURE 4.2 Illustration of the principle of optimality.

The Hamilton-Jacobi Equation

Now, we shall derive a partial differential Equation—the **Hamilton-Jacobi Equation**—which is then used as a **necessary condition** for the control to be **optimal** in the sense of minimum J, i.e., the optimal value of performance index J, denoted by J^*.

Consider a plant described by

$$\mathbf{x}(t) = f(\mathbf{x}(t), \mathbf{u}(t), t). \tag{4.26}$$

We are required to find optimal control law $u^*(t)$, so that the quadratic performance index,

$$J(\mathbf{x}(t), \mathbf{u}(\tau), t) = \psi(\mathbf{x}(t_f), t_f) + \int_t^{t_f} L(\mathbf{x}(\tau), \mathbf{u}(\tau), \tau) d\tau \tag{4.27}$$

$$t \leq \tau \leq t_f$$

$(t \leq \tau \leq t_f)$; is minimized, subject to the initial condition $\mathbf{x}(t)$.

The optimal value of the performance index is given by

$$J^*(\mathbf{x}(t), t) = J(\mathbf{x}(t), \mathbf{u}^*(\tau), t). \tag{4.28}$$

$$t \leq \tau \leq t_f$$

We may also write

$$J^*(\mathbf{x}(t), t) = \min_{\substack{\mathbf{u}(\tau) \\ t \leq \tau \leq t_f}} \left\{ \psi(\mathbf{x}(t_f), t_f) + \int_t^{t_f} L(\mathbf{x}(\tau), \mathbf{u}(\tau), \tau) \, d\tau \right\}. \tag{4.29}$$

Subdividing the interval, we get

$$J^*(\mathbf{x}(t),t) = \min_{\substack{\mathbf{u}(\tau) \\ t \le \tau \le t_f}} \left\{ \psi(\mathbf{x}(t_f),t_f) + \int_t^{t+\Delta t} L(\mathbf{x}(\tau),\mathbf{u}(\tau),\tau)\,d\tau + \int_{t+\Delta t}^{t_f} L(\mathbf{x}(\tau),\mathbf{u}(\tau),\tau)\,d\tau \right\}$$

or,

$$J^*(\mathbf{x}(t),t) = \min_{\substack{\mathbf{u}(\tau) \\ t \le \tau \le t_f}} \left\{ \int_t^{t+\Delta t} L(\mathbf{x}(\tau),\mathbf{u}(\tau),\tau)\,d\tau + \left[\psi(\mathbf{x}(t_f),t_f) + \int_{t+\Delta t}^{t_f} L(\mathbf{x}(\tau),\mathbf{u}(\tau),\tau)\,d\tau \right] \right\}.$$

Using the **principle of optimality**, we may write

$$J^*(\mathbf{x}(t),t) = \min_{\substack{\mathbf{u}(\tau) \\ t \le \tau \le t+\Delta t}} \left\{ \int_t^{t+\Delta t} L(\mathbf{x}(\tau),\mathbf{u}(\tau),\tau)\,d\tau + J^*(\mathbf{x}(t+\Delta t),t+\Delta t) \right\} \qquad (4.30)$$

where, $J^*(\mathbf{x}(t+\Delta t),t+\Delta t)$ is the optimal value of the performance index for the time-interval $(t+\Delta t) \le \tau \le t_f$, with the initial condition,

$$\mathbf{x}(t+\Delta t) = \mathbf{x}(t) + \int_t^{t+\Delta t} f(\mathbf{x}(\tau),\mathbf{u}(\tau),\tau)\,d\tau$$

$$= \mathbf{x}(t) + \Delta \mathbf{x}.$$

Note that Equation (4.30) is an **expression** of the **Principle of Optimality** that was illustrated in Figure 4.2.

Expanding $J^*(\mathbf{x}(t+\Delta t),t+\Delta t)$ in Taylor's series about the point $(\mathbf{x}(t),t)$, assuming that J^* has continuous first and second partial derivatives (neglecting higher order terms), we have

$$J^*(\mathbf{x}(t),t) = \min_{\substack{\mathbf{u}(\tau) \\ t \le \tau \le t+\Delta t}} \left\{ \int_t^{t+\Delta t} L(\mathbf{x}(\tau),\mathbf{u}(\tau),\tau)\,d\tau + J^*(\mathbf{x}(t)+t) \right.$$

$$\left. + \left[\frac{\partial J^*(\mathbf{x}(t),t)}{\partial t} \right] \Delta t + \left[\frac{\partial J^*(\mathbf{x}(t),t)}{\partial \mathbf{x}} \right]^T \Delta \mathbf{x} \right\}$$

or,

$$J^*(\mathbf{x}(t),t) = J^*(\mathbf{x}(t),t) + \left[\frac{\partial J^*(\mathbf{x}(t),t)}{\partial t} \right] \Delta t$$

$$+ \min_{\substack{\mathbf{u}(\tau) \\ t \le \tau \le t+\Delta t}} \left\{ \int_t^{t+\Delta t} L(\mathbf{x}(\tau),\mathbf{u}(\tau),\tau)\,d\tau + \left[\frac{\partial J^*(\mathbf{x}(t),t)}{\partial \mathbf{x}} \right]^T \Delta \mathbf{x} \right\}$$

or, $\quad 0 = \left[\dfrac{\partial J^*(\mathbf{x}(t), t)}{\partial t} \right] \Delta t + \min_{\substack{u(\tau) \\ t \le \tau \le t + \Delta t}} \left\{ \displaystyle\int_t^{t + \Delta t} L(\mathbf{x}(\tau), \mathbf{u}(\tau), \tau) d\tau + \left[\dfrac{\partial J^*(\mathbf{x}(t), t)}{\partial \mathbf{x}} \right]^T \Delta \mathbf{x} \right\}.$ (4.31)

Dividing both sides of Equation (4.31) by Δt, taking the limit $\Delta t \to 0$ and using the following two results :

1. $\displaystyle\lim_{\Delta t \to 0} \frac{\Delta \mathbf{x}}{\Delta t} = \lim_{\Delta t \to 0} \frac{\mathbf{x}(t + \Delta t) - \mathbf{x}(t)}{\Delta t} = \mathbf{x}(t) = f(\mathbf{x}(t), \mathbf{u}(t), t)$

2. $\displaystyle\lim_{\Delta t \to 0} \left[\frac{1}{\Delta t} \int_t^{t + \Delta t} L(\mathbf{x}(\tau), \mathbf{u}(\tau), \tau) d\tau \right] = L(\mathbf{x}(t), \mathbf{u}(t), t)$

We obtain

$$-\frac{\partial J^*(\mathbf{x}(t), t)}{\partial t} = \min_{\mathbf{u}(t)} \left\{ L(\mathbf{x}(t), \mathbf{u}(t), t) + \left[\frac{\partial J^*(\mathbf{x}(t), t)}{\partial \mathbf{x}} \right]^T f(\mathbf{x}(t), \mathbf{u}(t), t) \right\} \quad (4.32)$$

with the boundary condition that

$$J^*(\mathbf{x}(t_f), t_f) = \psi(\mathbf{x}(t_f), t_f).$$

Equation (4.32) is the **first statement** of the **Hamilton Jacobi Equation**.

Now, define the **Hamiltonian** $\mathcal{H}$ as :

$$\mathcal{H} \lim_{x \to \infty} \sqrt{a^2 + b^2} \left(\mathbf{x}(t), \mathbf{u}(t), \frac{\partial J^*}{\partial \mathbf{x}}, t \right) = L(\mathbf{x}(t), \mathbf{u}(t), t) + \left[\frac{\partial J^*(\mathbf{x}(t), t)}{\partial \mathbf{x}} \right]^T. \quad (4.33)$$

Using the preceeding definition of Hamiltonian, we can rewrite Equation (4.32) as

$$-\frac{\partial J^*(\mathbf{x}(t), t)}{\partial t} = \min_{\mathbf{u}(t)} \left\{ \mathcal{H} \left(\mathbf{x}(t), \mathbf{u}(t), \frac{\partial J^*}{\partial \mathbf{x}}, t \right) \right\} \quad (4.34)$$

Thus, the optimal control is obtained by minimizing the Hamiltonian. The optimal value of the Hamiltonian is given by

$$\mathcal{H}^*\left(\mathbf{x}(t), \frac{\partial J^*}{\partial \mathbf{x}}, t\right) = \mathcal{H}\left(\mathbf{x}(t), \mathbf{u}^*(t), \frac{\partial J^*}{\partial \mathbf{x}}, t\right). \tag{4.35}$$

We can rewrite Equation (4.34) as

$$\frac{\partial J^*(\mathbf{x}(t), t)}{\partial t} = -\mathcal{H}^*\left(\mathbf{x}(t), \frac{\partial J^*}{\partial \mathbf{x}}, t\right). \tag{4.36}$$

Equation (4.32) is the **second statement** of the **Hamilton-Jacobi equation**. The solution of the Hamilton-Jacobi equation gives a **candidate** for optimal control. It is most often used **as a check** on the optimality of a control derived from the minimum principle. Moreover, the Hamilton-Jacobi equation provides us a **bridge** from the **dynamic programming approach** to **variational methods** as we may formally obtain the minimum principle by taking the appropriate partial derivatives of the Hamilton-Jacobi equation. However, the minimum principle derived from the Hamilton-Jacobi equation may not be applicable to a **class** of optimal control problems.

4.6   SOLUTIONS TO OPTIMAL CONTROL PROBLEMS

In this section, we will solve various optimal control problems, already stated in Section (4.3), in order to deduce the optimal performance index J^* and associated optimal control for those optimal control problems.

4.6.1   Finite-Time State-Regulator Problem

Consider the optimal control problem given in Section 4.3.1. We are required to find the optimal control law $\mathbf{u}^*(t)$ so that the quadratic performance index given by (4.15) is minimized. From section 4.5, we know that the **necessary condition** of optimal control is that the Hamilton-Jacobi equation must be satisfied. Thus, for finite-time state-regulator problem, the Hamilton-Jacobi Equation (4.32) can be rewritten as

$$-\frac{\partial J^*(\mathbf{x}(t), t)}{\partial t} = \min_{\mathbf{u}(t)} \left\{ \frac{1}{2}\left[\mathbf{x}^T(t)\mathbf{Q}\mathbf{x}(t) + \mathbf{u}^T(t)\mathbf{R}\mathbf{u}(t)\right] + \left[\frac{\partial J^*(\mathbf{x}(t), t)}{\partial \mathbf{x}}\right]^T [\mathbf{A}\mathbf{x}(t) + \mathbf{B}\mathbf{u}(t)] \right\} \tag{4.37}$$

with the boundary condition that

$$J^*(\mathbf{x}(t_f), t_f) = \frac{1}{2}\mathbf{x}^T(t_f)\,\mathbf{H}\mathbf{x}(t_f). \tag{4.38}$$

In order to carry out the **minimization process**, we differentiate both sides of Equation (4.37) with respect to $\mathbf{u}(t)$. We get

$$0 = \mathbf{R}\mathbf{u}^*(t) + \mathbf{B}^T\left[\frac{\partial J^*(\mathbf{x}(t), t)}{\partial \mathbf{x}}\right]. \tag{4.39}$$

Since, the second derivative of $\left[-\dfrac{\partial J^*(\mathbf{x}(t), t)}{\partial \mathbf{x}}\right]$ with respect to $\mathbf{u}(t)$, gives a positive-definite matrix $\mathbf{R}$. Therefore, the control law $\mathbf{u}^*(t)$, given by Equation (4.39) is the optimal control due to the fact that **it satisfies the necessary and sufficient conditions for minimum**.

Solving Equation (4.39) for u*(t), we get

$$\mathbf{u}^*(t) = -\,\mathbf{R}^{-1}\mathbf{B}^T\left[\frac{\partial J^*(\mathbf{x}(t), t)}{\partial \mathbf{x}}\right]. \tag{4.40}$$

Substituting the value of $\mathbf{u}^*(t)$ into Equation (4.37), we obtain

$$-\frac{\partial J^*(\mathbf{x}(t), t)}{\partial t} = \frac{1}{2}\mathbf{x}^T(t)\,\mathbf{Q}\mathbf{x}(t) - \frac{1}{2}\left[\frac{\partial J^*(\mathbf{x}(t), t)}{\partial \mathbf{x}}\right]^T\mathbf{B}\mathbf{R}^{-1}\mathbf{B}^T\left[\frac{\partial J^*(\mathbf{x}(t), t)}{\partial \mathbf{x}}\right] + \left[\frac{\partial J^*(\mathbf{x}(t), t)}{\partial \mathbf{x}}\right]^T\mathbf{A}\mathbf{x}(t). \tag{4.41}$$

Since, the optimal value of the performance index is a time-varying quadratic function of $\mathbf{x}(t)$. Therefore, we may write

$$J^*(\mathbf{x}(t), t) = \frac{1}{2}\mathbf{x}^T(t)\,\mathbf{P}(t)\,\mathbf{x}(t) \tag{4.42}$$

where $\mathbf{P}(t)$ is a real, symmetric, positive-definite matrix for $t < t_f$.

Comparing Equations (4.42) and (4.38), we get

$$\mathbf{P}(t_f) = \mathbf{H}. \tag{4.43}$$

Now, substituting Equation (4.42) in Equation (4.41), we obtain

$$-\frac{1}{2}\mathbf{x}^T(t)\,\frac{d\mathbf{P}(t)}{dt}\mathbf{x}(t) = \frac{1}{2}\mathbf{x}^T(t)\,\mathbf{Q}\mathbf{x}(t) - \frac{1}{2}\mathbf{x}^T(t)\,\mathbf{P}(t)\,\mathbf{B}\mathbf{R}^{-1}\mathbf{B}^T\mathbf{P}(t)\,\mathbf{x}(t) + \mathbf{x}^T(t)\,\mathbf{P}(t)\,\mathbf{A}\mathbf{x}(t).$$

On simplification, we have

$$\mathbf{x}^T(t)\,[\mathbf{P}(t) + \mathbf{Q} - \mathbf{P}(t)\,\mathbf{B}\mathbf{R}^{-1}\mathbf{B}^T\mathbf{P}(t) + 2\mathbf{P}(t)\,\mathbf{A}]\mathbf{x}(t) = 0. \qquad (4.44)$$

We know that in the scalar function $\mathbf{z}^T\mathbf{W}\mathbf{z}$, only the symmetric part of the matrix $\mathbf{W}$, given by

$$\mathbf{W}_{\text{symmetric}} = \frac{\mathbf{W} + \mathbf{W}^T}{2},$$

is of importance.

If we examine the Equation (4.44), we find that all the terms within brackets are already symmetric except the last term.

So, the symmetric part of $\quad [2\mathbf{P}(t)\mathbf{A}] = 2\left[\dfrac{\mathbf{P}(t)\mathbf{A} + \mathbf{A}^T\mathbf{P}(t)}{2}\right]$

$$= \mathbf{P}(t)\mathbf{A} + \mathbf{A}^T\mathbf{P}(t).$$

Therefore, in order for Equation (4.44) to be satisfied for an arbitrary state $\mathbf{x}(t)$, it is necessary that the matrix differential Equation,

$$\mathbf{P}(t) + \mathbf{Q} - \mathbf{P}(t)\mathbf{B}\mathbf{R}^{-1}\mathbf{B}^T\mathbf{P}(t) + \mathbf{P}(t)\mathbf{A} + \mathbf{A}^T\mathbf{P}(t) = 0 \qquad (4.45)$$

is satisfied, subject to the boundary condition given by Equation (4.43).

Equation (4.45) is referred to as the **Matrix Riccati Equation**. The matrix $\mathbf{P}(t)$ is sometimes referred to as the **Riccati Gain Matrix**. Once the matrix $\mathbf{P}(t)$ is determined for $t_o \leq t \leq t_f$, the optimal control law $u^*(t)$ can be obtained from Equations (4.40) and (4.42), as follows :

$$\mathbf{u}^*(t) = -\mathbf{R}^{-1}\mathbf{B}^T\mathbf{P}(t)\,\mathbf{x}(t) = \mathbf{K}(t)\,\mathbf{x}(t) \qquad (4.46)$$

where,

$$\mathbf{K}(t) = -\mathbf{R}^{-1}\mathbf{B}^T\mathbf{P}(t).$$

We shall now **summarize** the results of what we have just discussed in this section, as follows:

For the plant, described by (4.14) and the performance index (4.15), a **unique** optimal control exists and is given by

$$\mathbf{u}^*(t) = \mathbf{K}(t)\,\mathbf{x}(t);\ \mathbf{K}(t) = -\mathbf{R}^{-1}\mathbf{B}^T\mathbf{P}(t)$$

where $\mathbf{P}(t)$ is a real, symmetric, positive-definite matrix, which is the **solution** of the Matrix Riccati Equation

$$\mathbf{P}(t) + \mathbf{Q} - \mathbf{P}(t)\mathbf{B}\mathbf{R}^{-1}\mathbf{B}^T\mathbf{P}(t) + \mathbf{P}(t)\mathbf{A} + \mathbf{A}^T\mathbf{P}(t) = 0$$

with the boundary condition

$$\mathbf{P}(t_f) = \mathbf{H}.$$

The optimal value of performance index is given by

$$J^* = \frac{1}{2}\mathbf{x}^T(t)\,\mathbf{P}(t)\,\mathbf{x}(t).$$

4.6.2   Infinite-Time State-Regulator Problem

Consider the optimal control problem given in Section 4.3.2. We are required to find the optimal control law $\mathbf{u}^*(t)$ so that the quadratic performance index given by (4.18) is minimized. It has already been stated that for infinite-time state-regulator problem, the **terminal constraints have no significance** in performance index J (therefore $\mathbf{H} = \mathbf{0}$), and the solution of this optimal control problem exists **only if** the plant described by (4.17) is **completely controllable**.

Hence, consider the Matrix Reccati Equation

$$\mathbf{P}(t) + \mathbf{Q} - \mathbf{P}(t)\mathbf{B}\mathbf{R}^{-1}\mathbf{B}^T\mathbf{P}(t) + \mathbf{P}(t)\mathbf{A} + \mathbf{A}^T\mathbf{P}(t) = 0 \tag{4.47}$$

with the boundary condition

$$\mathbf{P}(t_f) = \mathbf{H} = 0.$$

We shall solve the matrix Riccati Equation (4.47) **backward in time** with t_f as the starting time and $\mathbf{P}(t_f) = 0$ as the initial condition, and utilizing the fact that if the plant described by (4.17) is completely controllable, then $\lim_{t \to \infty} \mathbf{p}(t)$ tends to a unique positive-definite, constant matrix $\tilde{\mathbf{P}}$, i.e.,

$$\text{As } t_f \to \infty, \mathbf{P}(t) \to \mathbf{P} \text{ for all } t.$$

As $\tilde{\mathbf{P}}$ is a constant matrix, its time derivative is zero. Now substituting this result into the Matrix Riccati Equation (4.47), we obtain

$$\mathbf{Q} - \mathbf{PBR}^{-1}\mathbf{B}^T\mathbf{P} + \mathbf{PA} + \mathbf{A}^T\mathbf{P} = 0. \tag{4.48}$$

which is often referred to as the **Reduced Matrix Riccati Equation**.

Unfortunately, the solution of Equation (4.48) is **not unique**. The desired unique answer is obtained by enforcing the requirement that $\tilde{\mathbf{P}}$ be positive-definite.

Note that, throughout the preceeding discussion, it has been assumed that the controlled plant (4.17) is **stable** so that

$$\lim_{t \to \infty} \mathbf{x}(t) = 0.$$

However, the plant described by (4.17) **may not always stable**.

The optimal control law in the infinite-time state-regulator problem, which directly follows from the results of Section 4.6.1, is given by

$$\mathbf{u}^*(t) = -\mathbf{R}^{-1}\mathbf{B}^T\tilde{\mathbf{P}}\mathbf{x}(t) = \tilde{\mathbf{K}}\mathbf{x}(t) \tag{4.49}$$

where

$$\mathbf{K} = -\mathbf{R}^{-1}\mathbf{B}^T\mathbf{P}$$

The resulting optimal system is, thus, a **linear time-invariant system**. The optimal value of the performance index is obtained as follows :
From Equation (4.42), we have

$$J^*(\mathbf{x}(t), t) = \frac{1}{2}\mathbf{x}^T(t) \, \mathbf{P}(t) \, \mathbf{x}(t).$$

For the infinite-time case, we may write

$$J^*(\mathbf{x}(t), t) = \frac{1}{2}\int_\infty^{t_0} \left\{ \frac{d}{dt}[\mathbf{x}^T(t) \, \tilde{\mathbf{P}}\mathbf{x}(t)] \, dt \right\}$$

$$= \left[\frac{1}{2} \, \mathbf{x}^T(t) \, \tilde{\mathbf{P}}\mathbf{x}(t) \right]_\infty^{t_0}$$

$$= \frac{1}{2}\mathbf{x}^T(t_0) \, \tilde{\mathbf{P}}\mathbf{x}(t_0).$$

Note that, $\lim_{t \to \infty} \mathbf{x}(t) \to 0$, as per the stability assumption made earlier.

We shall now **summarise** the results of what we have just discussed in this section, as follows:

For the plant, described by (4.17) and the performance index (4.18), a **unique** optimal control exists and is given by

$$\mathbf{u}^*(t) = \tilde{\mathbf{K}}\mathbf{x}(t); \ \tilde{\mathbf{K}} = -\mathbf{R}^{-1}\mathbf{B}^T\tilde{\mathbf{P}}$$

where $\tilde{\mathbf{P}}$ is a unique, positive-definite, constant matrix, and is the solution of the Reduced Matrix Riccati Equation

$$\mathbf{Q} - \mathbf{P}\mathbf{B}\mathbf{R}^{-1}\mathbf{B}^T\mathbf{P} + \mathbf{P}\mathbf{A} + \mathbf{A}^T\mathbf{P} = 0.$$

The optimal value of the performance index is given by

$$J^* = \frac{1}{2} \mathbf{x}^T(t_0)\tilde{\mathbf{P}}\mathbf{x}(t_0).$$

4.6.3   The Output-Regulator Problem

Consider the optimal control problem given in Section 4.3.3. We are required to find the optimal control law $\mathbf{u}^*(t)$ so that the quadratic performance index given by (4.20) is minimized. It has already been stated that if the plant described by (4.19) is **observable,** then we can reduce the output regulator problem to the state regulator problem in order to find the solution to the output regulator problem as follows :

Substitute $\mathbf{y}(t) = \mathbf{C}\mathbf{x}(t)$ in Equation (4.20) and we obtain

$$J = \frac{1}{2} \mathbf{x}^T(t_f) \mathbf{C}^T\mathbf{H}\mathbf{C} \, \mathbf{x}(t_f) + \frac{1}{2} \int_{t_0}^{t_f} [\mathbf{x}^T(t) \, \mathbf{C}^T\mathbf{Q}\mathbf{C} \, \mathbf{x}(t) + \mathbf{u}^T(t)\mathbf{R}\mathbf{u}(t)] \, dt. \qquad (4.50)$$

Comparing Equation (4.50) with Equation (4.15), we can immediately apply the results of the finite-time state-regulator problem for the output-regulator problem under consideration. We shall now **summarize** the results as follows:

For the plant, described by (4.19) and the performance index (4.20), a **unique** optimal control exists and is given by

$$\mathbf{u}^*(t) = \mathbf{K}(t) \, \mathbf{x}(t); \ \mathbf{K}(t) = -\mathbf{R}^{-1}\mathbf{B}^T\mathbf{P}(t)$$

where $\mathbf{P}(t)$ is a real, symmetric, positive-definite matrix, and is the solution of the Matrix Riccati Equation

$$\mathbf{P}(t) + \mathbf{C}^T\mathbf{Q}\mathbf{C} - \mathbf{P}(t)\,\mathbf{B}\mathbf{R}^{-1}\mathbf{B}^T\mathbf{P}(t) + \mathbf{P}(t)\,\mathbf{A} + \mathbf{A}^T\mathbf{P}(t) = 0 \qquad (4.51)$$

with the boundary condition

$$\mathbf{P}(t_f) = \mathbf{C}^T\mathbf{H}\mathbf{C}. \qquad (4.52)$$

Note that the results for the output-regulator problem for the case when final time t_f is not constrained, can easily be obtained from the results of Section 4.6.2.

4.6.4   The Tracking-Problem

Consider the optimal control problem given in Section 4.3.4. We are required to find the optimal control law $\mathbf{u}^*(t)$ so that the quadratic performance index given by (4.23) is minimized. If we consider only those desired outputs $\mathbf{r}(t)$, that could be generated by arbitrary initial conditions $\mathbf{z}(t_o)$ in the system, described by

$$\mathbf{z}(t) = \mathbf{A}\mathbf{z}(t) \qquad (4.53\ A)$$
$$\mathbf{r}(t) = \mathbf{C}\mathbf{z}(t) \qquad (4.53\ B)$$

where matrices $\mathbf{A}$ and $\mathbf{C}$ are same as those of the plant (4.21), then we can reduce the tracking problem to the output regulator problem in order to find the solution to the tracking problem as follows:

Define a new state-variable

$$\boldsymbol{\omega}(t) = \mathbf{x}(t) - \mathbf{z}(t)$$

so that $$\boldsymbol{\omega}(t) = \mathbf{A}\boldsymbol{\omega}(t) + \mathbf{B}\mathbf{u}(t) \qquad (4.54\ A)$$
$$\mathbf{e}(t) = \mathbf{C}\boldsymbol{\omega}(t) \qquad (4.54\ B)$$

where $\mathbf{e}(t)$ is the **tracking error** and is given by Equation (4.22).

Now, we can easily apply the results of the output-regulator problem for the tracking problem under consideration; which immediately gives the optimal control for this tracking problem.

We can now **summarize** the results as follows :

For the plant, described by (4.21) and the performance index (4.23), a unique optimal control exists and is given by

$$\mathbf{u}^*(t) = \mathbf{K}(t)\,[\mathbf{x}(t) - \mathbf{z}(t)]; \; \mathbf{K}(t) = -\mathbf{R}^{-1}\mathbf{B}^T\mathbf{P}(t)$$

where P(t) is a real, symmetric, positive-definite matrix, which is the solution of the Matrix Riccati Equation

$$\mathbf{P}(t) + \mathbf{C}^T\mathbf{Q}\mathbf{C} - \mathbf{P}(t)\,\mathbf{B}\mathbf{R}^{-1}\mathbf{B}^T\mathbf{P}(t) + \mathbf{P}(t)\mathbf{A} + \mathbf{A}^T\mathbf{P}(t) = 0 \qquad (4.55)$$

with the boundary condition

$$\mathbf{P}(t_f) = \mathbf{C}^T\mathbf{H}\mathbf{C}. \qquad (4.56)$$

Note that the results for the tracking problem for the case when final time t_f is not constrained, can easily be obtained from the results of Section 4.6.2.

Example 4.1. *A system is governed by the state-Equations:*

$$\mathbf{x} = \begin{bmatrix} 0 & 1 \\ 0 & 0 \end{bmatrix}\mathbf{x} + \begin{bmatrix} 0 \\ 1 \end{bmatrix}u. \qquad (4.57)$$

Find the control law which minimizes the performance index,

$$J = \int_0^\infty (x_1^2 + u^2)\, dt. \qquad (4.58)$$

Solution. From Equation (4.57), we have

$$\mathbf{A} = \begin{bmatrix} 0 & 1 \\ 0 & 0 \end{bmatrix} \text{ and } \mathbf{B} = \begin{bmatrix} 0 \\ 1 \end{bmatrix}.$$

As the given problem is an infinite-time state-regulator problem, so comparing (4.58) with the performance index given by (4.18), we have

$$\mathbf{Q} = 2\begin{bmatrix} 1 & 0 \\ 0 & 0 \end{bmatrix} = \begin{bmatrix} 2 & 0 \\ 0 & 0 \end{bmatrix}$$

and, $\mathbf{R} = 2[1] = [2].$

Now, the Reduced Matrix Ricaati Equation is given by

$$\mathbf{Q} - \mathbf{PBR^{-1}B^{T}P} + \mathbf{PA} + \mathbf{A^{T}P} = 0.$$ (4.59)

Putting all values in Equation (4.59), we get

$$\begin{bmatrix} 2 & 0 \\ 0 & 0 \end{bmatrix} - \begin{bmatrix} p_{11} & p_{12} \\ p_{21} & p_{22} \end{bmatrix}\begin{bmatrix} 0 \\ 1 \end{bmatrix}[1/2][0 \quad 1]\begin{bmatrix} p_{11} & p_{12} \\ p_{21} & p_{22} \end{bmatrix}$$
$$+ \begin{bmatrix} p_{11} & p_{12} \\ p_{21} & p_{22} \end{bmatrix}\begin{bmatrix} 0 & 1 \\ 0 & 0 \end{bmatrix} + \begin{bmatrix} 0 & 0 \\ 1 & 0 \end{bmatrix}\begin{bmatrix} p_{11} & p_{12} \\ p_{21} & p_{22} \end{bmatrix} = \begin{bmatrix} 0 & 0 \\ 0 & 0 \end{bmatrix}.$$

We assume that $\tilde{\mathbf{P}}$ is symmetric, we may take

$$p_{12} = p_{21}.$$

Thus,

$$\begin{bmatrix} 2 & 0 \\ 0 & 0 \end{bmatrix} - \begin{bmatrix} p_{11} & p_{12} \\ p_{12} & p_{22} \end{bmatrix}\begin{bmatrix} 0 \\ 1 \end{bmatrix}[1/2][0 \quad 1]\begin{bmatrix} p_{11} & p_{12} \\ p_{12} & p_{22} \end{bmatrix}$$
$$+ \begin{bmatrix} p_{11} & p_{12} \\ p_{12} & p_{22} \end{bmatrix}\begin{bmatrix} 0 & 1 \\ 0 & 0 \end{bmatrix} + \begin{bmatrix} 0 & 0 \\ 1 & 0 \end{bmatrix}\begin{bmatrix} p_{11} & p_{12} \\ p_{12} & p_{22} \end{bmatrix} = \begin{bmatrix} 0 & 0 \\ 0 & 0 \end{bmatrix}.$$

On simplification, we obtain

$$\left.\begin{aligned} -\frac{p_{12}^2}{2} + 2 &= 0 \\ p_{11} - \frac{p_{12}p_{22}}{2} &= 0 \\ -\frac{p_{22}^2}{2} + 2p_{12} &= 0 \end{aligned}\right\}.$$ (4.60)

On solving Equations (4.60), we have

$$p_{11} = 2\sqrt{2},$$
$$p_{12} = 2,$$
and,
$$p_{22} = 2\sqrt{2}.$$

Thus,

$$\mathbf{P} = \begin{bmatrix} p_{11} & p_{12} \\ p_{12} & p_{22} \end{bmatrix} = \begin{bmatrix} 2\sqrt{2} & 2 \\ 2 & 2\sqrt{2} \end{bmatrix}.$$

Now, from Equation (4.49), the optimal control law is given by

$$\mathbf{u}^{*}(t) = -\mathbf{R}^{-1}\mathbf{B}^{T}\tilde{\mathbf{P}}\mathbf{x}(t) = \tilde{\mathbf{K}}\mathbf{x}(t).$$

Putting all the values, we have

$$\mathbf{u}^{*}(t) = -[1/2][0 \quad 1]\begin{bmatrix} 2\sqrt{2} & 2 \\ 2 & 2\sqrt{2} \end{bmatrix}\begin{bmatrix} x_{1}(t) \\ x_{2}(t) \end{bmatrix}$$

or, $$\mathbf{u}^{*}(t) = -x_{1}(t) - \sqrt{2}x_{2}(t)$$

Example 4.2. *The following differential Equation describes a first-order system:*

$$x(t) = 2x(t) + u(t).$$

Obtain the optimal control law, so as to minimize the performance index,

$$J = \frac{1}{2}\int_{0}^{t_{f}}\left(3x^{2} + \frac{1}{4}u^{2}\right)dt$$

and $$t_{f} = 1\,sec.$$

Solution. The problem is a finite-time state-regulator problem. As the system is of the first-order, for this particular problem, the matrices A, B, H, Q and R reduce to scalars a, b, h, q, and r respectively ; where

$$a = 2,$$
$$b = 1,$$
$$h = 0,$$
$$q = 3,$$
$$r = \frac{1}{4}.$$

Moreover, the matrix $\mathbf{P}(t)$ also reduces to a scalar function of time $p(t)$.

Now, the Matrix Ricaati Equation (4.45) reduces to the scalar differential equation and is given by

$$p(t) + q - \frac{b^2}{r} p^2(t) + 2ap(t) = 0 \qquad (4.61)$$

with the boundary condition,

$$p(t_f) = h = 0.$$

Putting all values in Equation (4.61), we have

$$p(t) + 3 - 4p^2(t) + 4p(t) = 0$$

or, $$\frac{dp(t)}{dt} + 3 - 4p^2(t) + 4p(t) = 0$$

or, $$\frac{dp(t)}{dt} = 4p^2(t) - 4p(t) - 3$$

or, $$\frac{dp(t)}{dt} = 4p^2(t) - 6p(t) + 2p(t) - 3$$

or, $$\frac{dp(t)}{dt} = 2p(t)\{2p(t) - 3\} + \{2p(t) - 3\}$$

or, $$\frac{dp(t)}{dt} = \{2p(t) - 3\}\{2p(t) + 1\}$$

or, $$\frac{dp(t)}{44\left\{p(t) - \frac{3}{2}\right\}\left\{p(t) + \frac{1}{2}\right\}} = dt.$$

On integration, we obtain

$$\int_{t_f}^{t} \frac{dp(\tau)}{4\left\{p(\tau) - \frac{3}{2}\right\}\left\{p(\tau) + \frac{1}{2}\right\}} = \int_{t_f}^{t} d\tau$$

or, $$\frac{1}{8}\int_{t_f}^{t} \left\{\frac{1}{\left(p(\tau) - \frac{3}{2}\right)} - \frac{1}{\left(p(\tau) + \frac{1}{2}\right)}\right\} dp(\tau) = (t - t_f)$$

or,
$$\frac{1}{8}\left\{\ln\left(p(\tau)-\frac{3}{2}\right)-\ln\left(p(\tau)+\frac{1}{2}\right)\right\}_{t_f}^{t}=(t-t_f)$$

or,
$$\frac{1}{8}\left\{\ln\left(\frac{p(t)-\frac{3}{2}}{p(t)+\frac{1}{2}}\right)-\ln\left(\frac{p(t_f)-\frac{3}{2}}{p(t_f)+\frac{1}{2}}\right)\right\}=(t-t_f)$$

or,
$$\frac{1}{8}\ln\left\{\frac{\left(p(t)-\frac{3}{2}\right)\left(p(t_f)+\frac{1}{2}\right)}{\left(p(t)+\frac{1}{2}\right)\left(p(t_f)-\frac{3}{2}\right)}\right\}=(t-t_f)$$

or,
$$\frac{\left(p(t)-\frac{3}{2}\right)\left(p(t_f)+\frac{1}{2}\right)}{\left(p(t)+\frac{1}{2}\right)\left(p(t_f)-\frac{3}{2}\right)}=\exp\left\{8(t-t_f)\right\}.$$

Put $p(t_f) = 0$, and we get

$$\frac{\frac{1}{2}\left(p(t)-\frac{3}{2}\right)}{-\frac{3}{2}\left(p(t)+\frac{1}{2}\right)}=\exp\left\{8(t-t_f)\right\}$$

or,
$$p(t)-\frac{3}{2}=-3\left(p(t)+\frac{1}{2}\right)\exp\left\{8(t-t_f)\right\}$$

or,
$$p(t)\left[1+3\exp\left\{8(t-t_f)\right\}\right]=\frac{3}{2}\left[1-\exp\left\{8(t-t_f)\right\}\right]$$

or,
$$p(t)=\frac{\frac{3}{2}\left[1-\exp\left\{8(t-t_f)\right\}\right]}{1+3\exp\left\{8(t-t_f)\right\}}.$$

Now, the optimal control law is given by

$$u^*(t) = -\frac{b}{r}p(t)\,x(t)$$

or, $$u^*(t) = -4p(t)\,x(t)$$

or, $$u^*(t) = -4\left\{\frac{\frac{3}{2}[1 - \exp\{8(t-1)\}]}{[1 + 3\exp\{8(t-1)\}]}\right\}x(t)$$ **Ans.**

In block-diagram representation, the optimal-system is shown in Figure 4.3 given here:

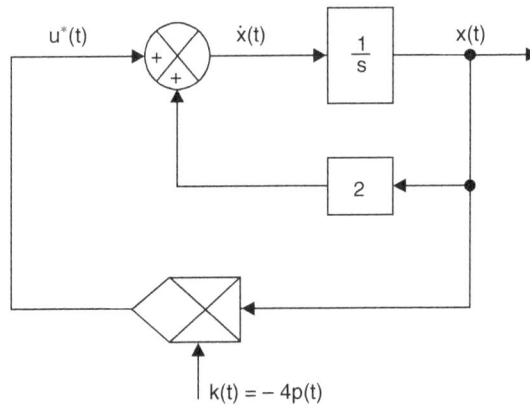

FIGURE 4.3 Block diagram representation of the optimal system of Example 4.2.

EXERCISES

1. A plant is governed by the Equations :

$$\begin{bmatrix} x_1 \\ x_2 \end{bmatrix} = \begin{bmatrix} 1 & 0 \\ -1 & 2 \end{bmatrix}\begin{bmatrix} x_1 \\ x_2 \end{bmatrix} + \begin{bmatrix} 1 \\ 0 \end{bmatrix}u.$$

(a) Prove that the system is unstable.

(b) Prove that the system is controllable.

(c) Select any values for matrices $\mathbf{Q}$ and $\mathbf{R}$ with the constraint that they are positive-definite and then design a controller for the plant so as to minimize the performance index given by

$$J = \frac{1}{2}\int_0^\infty (\mathbf{x}^T\mathbf{Q}\mathbf{x} + \mathbf{u}^T\mathbf{R}\mathbf{u})\, dx.$$

Check whether the resulting overall system is stable or not.

2. A system is described by :

$$x_1 = x_2,$$
$$x_2 = u,$$
$$y = x_1.$$

Obtain the control law that minimizes

$$J = \frac{1}{2}\int_0^\infty (y^2 + u^2)dt.$$

3. Obtain the optimal control law for the system described by

$$\mathbf{x} = \mathbf{A}\mathbf{x} + \mathbf{B}u$$

where $\quad \mathbf{A} = \begin{bmatrix} 0 & 1 \\ -2 & -3 \end{bmatrix}; \mathbf{B} = \begin{bmatrix} 0 \\ 1 \end{bmatrix}$

so as to minimize the following performance index :

$$J = \int_0^\infty (\mathbf{x}^T\mathbf{x} + u^2)dt.$$

REFERENCES

(1) Anderson B.D.O., Moore John B., *Optimal Control*, Prentice-Hall, Englewood Cliffs, New Jersey, 1989.

(2) Benjamin C. Kuo, *Automatic Control Systems*, Prentice-Hall, Englewood Cliffs, New Jersey, 3rd Edition, 1975.

(3) I.J. Nagrath, M. Gopal, *Control Systems Engineering*, New Age International Publishers, New Delhi, 4th Edition, 2005.

(4) Katsuhiko Ogata, *Modern Control Engineering*, Prentice-Hall, Englewood Cliffs, New Jersey, 1970.

(5) M. Gopal, *Modern Control System Theory*, New Age International Publishers, New Delhi, 2nd Edition, 1993.

5

ADAPTIVE CONTROL

5.1   INTRODUCTION

Roughly speaking, **"to adapt"** means to change a behavior to conform to new circumstances. The term adaptive system has a variety of specific meanings, but it usually implies that the system is capable of accommodating unpredictable disturbances, whether these disturbances arise within the system or external to it.

Adaptation is a **fundamental characteristic** of living organisms since they attempt to maintain **physiological equilibrium** in the midst of changing environmental conditions. An approach to the design of adaptive systems is then to consider the adaptive aspects of human or animal behavior and to develop systems that behave somewhat **analogously.**

An adaptive controller is thus a controller that can modify its behavior in response to changes in the dynamics of the plant and the character of the disturbances. The basic objective of an adaptive controller is to maintain a consistent performance of a system in the presence of uncertainty or unknown variation in the plant parameters, which may occur due to nonlinear actuators, changes in the operating conditions of the plant and nonsatisfactory disturbances acting on the plant.

Since ordinary feedback also attempts to reduce the effects of the disturbances and plant uncertainty, then immediately the question arises in one's mind that what is the difference between feedback control and adaptive control ? An **adaptive controller** is a controller with **adjustable parameters** and a **mechanism for adjusting the parameters**. An adaptive control system can be thought of as having two loops. One loop is a normal feedback with the plant (process) and the controller. The other loop is a **parameter adjustment loop**. The parameter adjustment loop is often slower than the normal feedback loop. A block-diagram of an adaptive system is shown in Figure 5.1.

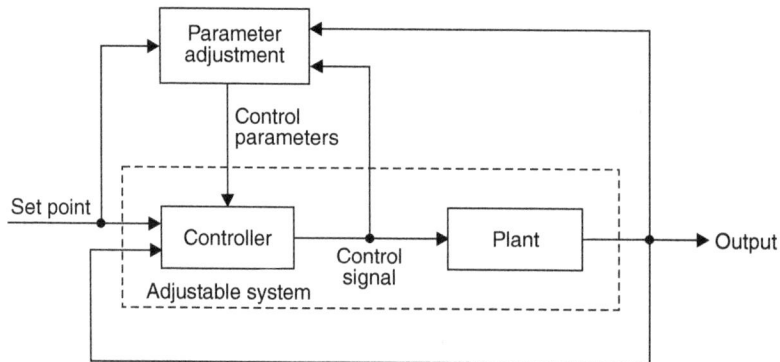

FIGURE 5.1 Block diagram of an adaptive system.

5.2   ESSENTIAL COMPONENTS OF AN ADAPTIVE SYSTEM

Three basic functions common to most of the adaptive systems are:
1. Identification of dynamic characteristics of the plant or measurement of an index of performance (IP).
2. Decision on control strategy based on the identification of dynamic characteristics of the plant or measurement of IP.
3. Online modification or actuation based on the decision made.

Depending on how these functions are brought about, we have different types of adaptive controllers. The essential components of an adaptive system are shown in Figure 5.2.

The **"adjustable system"** refers to a system whose performance can be adjusted either by modifying its parameters or by modifying its input signals. The identification of dynamic characteristics of the plant or measurement of an index of performance (IP) must be done continuously, or at least very

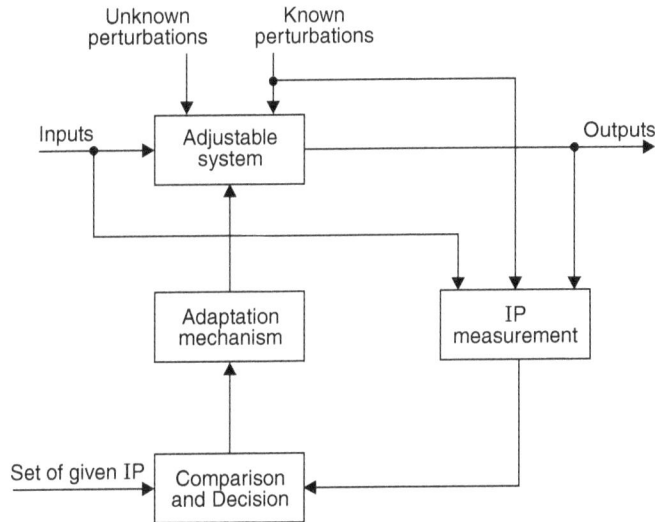

FIGURE 5.2 **Essential components of an adaptive system.**

frequently. There are several methods by which this can be accomplished. **"IP measurement"** may be done either directly or indirectly but without affecting the normal operation of the system. The **"comparison and decision"** block **"compares"** the set of given IP and the measured IP and then **"decides"** whether the measured values are within acceptable limits or not. If not, this block **"impels"** the **"adaptation mechanism"** to act appropriately to modify system performance, either by altering the parameters of the adjustable system, in which case the system will be called a **"parameter-adaptive control system"** or by changing the input signal, in which case the system is said to be a **"signal-synthesis adaptive control system."**

Now we will summarize what we have just stated in the form of a definition, given as :

"An adaptive system measures a certain index of performance (IP) using the inputs, the states, and the outputs of the adjustable system. From the comparison of the measured IP values and a set of given ones, the adaptation mechanism modifies the parameters of the adjustable system or generates an auxiliary input in order to maintain the IP values close to the set of given ones."

Figure 5.2 illustrates the definition. Note that it is not always possible for an adaptive system to be split into the distinct blocks as shown in Figure 5.2, as the systems are generally rather complicated.

5.3   ADAPTIVE SCHEMES

There are two principal approaches for designing adaptive controllers, namely, Model-Reference Adaptive Control (MRAC) systems and Self-Tuning Regulators (STR).

5.3.1   Model-Reference Adaptive Control (MRAC)

The Model-Reference Adaptive Control (MRAC) system was originally developed by Whitaker and his coworkers at MIT in 1938, for designing adaptive autopilots. In MRAC systems, the desired performance specifications are expressed in terms of a **reference model**, which gives the ideal response to a command signal. The aim is to make the output of an unknown plant approach **asymptotically** the output of a given reference model, which is the part of the control system shown in Figure 5.3.

The choice of the reference model has to satisfy two requirements :

1. It should reflect the performance specification in the control tasks such as rise-time, settling time, overshoot, or equivalent frequency domain characteristics.

2. The ideal behavior specified by the reference model should be achievable for the adaptive control system.

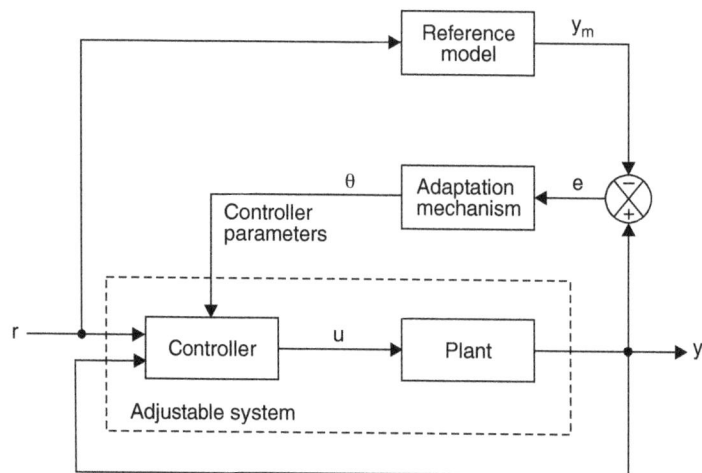

FIGURE 5.3 Block diagram of a model-reference adaptive
control (MRAC) system (parameter-adaptation).

The controller can be thought of as consisting of two loops. The inner loop is an ordinary feedback loop composed of the plant and the controller. The outer loop adjusts the controller parameters on the basis of the **tracking error** e, which is the difference between plant output y and model output y_m. This error is used by the adaptation mechanism to modify the controller parameters so as to minimize the tracking-error. This is a **parameter-adaptive** system (Figure 5.3).

One can also arrange the adaptation mechanism to generate an auxiliary input signal to minimize the tracking-error e. In this case the system will be a **signal-synthesis MRAC** (Figure 5.4).

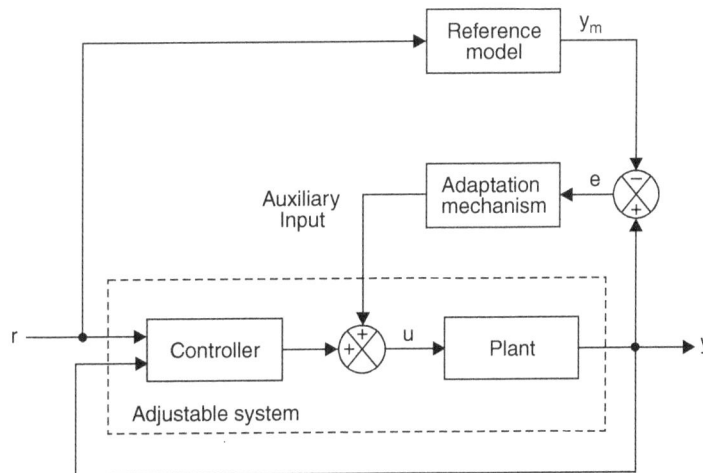

FIGURE 5.4   Block-Diagram of an MRAC system (signal-synthesis adaptation).

The **key problem** with MRAC is to design the adaptation mechanism to **guarantee** the stability of the control system as well as convergence of the tracking error to zero. This problem is **nontrivial**. An adaptive mechanism, called the **MIT Rule,** is the **original approach** to MRAC. The name is derived from the fact that it was developed at the Instrumentation Laboratory (now the Draper Laboratory) at MIT.

The MIT Rule

To present the MIT rule, we will consider that the controller in the closed-loop system has an adjustable parameter θ. The desired (ideal) closed-loop response specified by a reference model is y_m. One possibility is to adjust parameters in such a way that the **loss function**,

$$J(\theta) = \frac{1}{2}e^2 \tag{5.1}$$

is minimized. In Equation (5.1), e be the tracking error between the plant output y and the model output y_m. The loss function J can be minimized if we change the parameters in the direction of the **negative gradient** of J, generally known as **gradient descent approach**, in the following manner

$$\frac{d\theta}{dt} = -\gamma \frac{\partial J}{\partial \theta} = -\gamma e \frac{\partial e}{\partial \theta}. \tag{5.2}$$

Equation (5.2) is known as the **MIT rule.**

The quantity $\dfrac{\partial e}{\partial \theta}$, which is called the **sensitivity derivative** of the system, tells how the error is influenced by the adjustable parameter. If it is assumed that the parameter changes are slower than the other variables in the system, then the sensitivity derivative $\dfrac{\partial e}{\partial \theta}$ can be evaluated under the assumption that θ is constant. In practice it is necessary to make approximations to obtain the sensitivity derivative. The parameter γ determines the **adaptation rate**. The MIT rule can be regarded as a **gradient-scheme** to minimize the squared error e^2.

5.3.2   Self-Tuning Regulators (STR)

The self-tuning regulator (STR) is another important form of an adaptive system. A general architecture for the self-tuning control is given in Figure 5.5, which shows a block-diagram of a plant with a self-tuning regulator (STR).

The controller can be thought of as consisting of two loops. The inner loop is an ordinary feedback loop composed of the plant and the controller. The outer loop adjusts the controller parameter and is composed of a **recursive parameter estimator** and a **controller design criterion**. The controller of this construction is called a **self-tuning regulator** to emphasize that the controller **automatically tunes** its parameters to obtain the desired response of the closed-loop system.

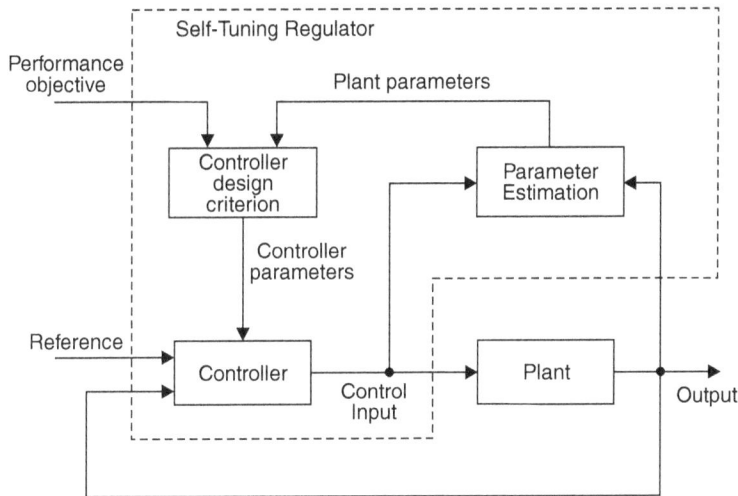

FIGURE 5.5   **Block-diagram of a self-tuning regulator (STR).**

Parameter estimation is a **key element** in a self-tuner and is performed **online**. The block labeled **"Parameter Estimation"** in Figure 5.5 gives an estimate of the plant parameters. The plant parameters are estimated based on the measurable plant input, plant output, and state signals. The estimates are then used in designing the controller as if they are equal to the true parameters (i.e., the uncertainties of the estimates are not considered). This is called the **certainty equivalence principle**.

The block labeled **"Controller Design Criterion"** in Figure 5.5 represents an **online solution** to a controller design problem for a system with known parameters. This design problem is called the **underlying design problem**. Such a problem can be associated with most adaptive control schemes, but it is often given indirectly. To evaluate adaptive control schemes, it is often useful to find the underlying design problem because it will give the characteristics of the system under the ideal conditions when the parameters are known exactly. The STR scheme is very flexible with respect to the choice of the underlying design and estimation methods.

The block labeled **"Controller"** in Figure 5.5, is an **implementation** of the controller whose parameters are obtained from the controller design criterion.

STR have become very popular in recent years because of their **versatility** and the **ease** with which they can be implemented with micro processors. The self-tuning adaptive controllers have the advantage of being able to adapt for any (**especially unmeasurable**) disturbances if designed well.

5.4   ABUSES OF ADAPTIVE CONTROL

An adaptive controller, being inherently **nonlinear**, is more complicated than a fixed-gain controller. Before attempting to use adaptive control, it is therefore, important to investigate whether the control problem might be solved by **constant-gain feedback**. There are many cases in which constant-gain feedback can perform **as good as** an adaptive controller. One way to proceed in deciding whether adaptive control should be used is illustrated in Figure 5.6.

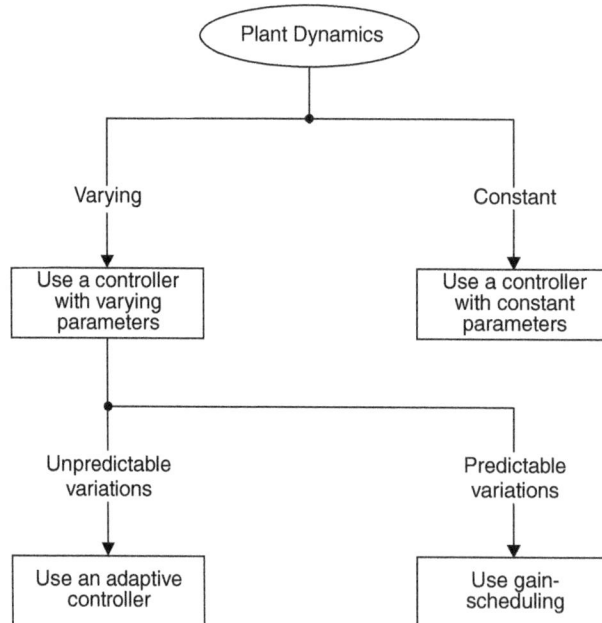

FIGURE 5.6   **Procedure to decide what type of controller to use.**

5.5   APPLICATIONS

There have been a number of applications of adaptive feedback control since the mid-1950s. The number of applications has increased drastically with the advent of the microprocessor, which has made the technology **cost-effective**.

Adaptive techniques have been used in regular industrial controllers since the early 1980s. Today, a large number of industrial control loops are under adaptive control. These include a wide range of applications in **aerospace**, **process control** (e.g., **temprature control** in a distillation column, **chemical reactor control**, **pulp dryer control**, **pulp digester control**, and **control of a rolling mill**, etc.), **ship steering**, **robotics**, **automobile**, and **biomedical systems**. Although there are many applications of adaptive control, yet **adaptive control is not a mature technology**.

EXERCISES

1. What is an adaptive control system ? Briefly discuss the essential aspects of an adaptive control system.

2. What are the various schemes of adaptive control ?

3. What are the situations in which adaptive control may be useful ? What factors would you consider when judging the need for adaptive control ?

4. List various applications of adaptive control ? What are the differences between feedback control and adaptive control ?

5. How can MRAC be employed using,

    (a) parameter adaptation,

    (b) signal-synthesis adaptation.

    Also, explain the MIT rule.

6. What is an underlying design problem ? How it is useful in the evaluation of an adaptive scheme ?

7. Look up the definitions of **"adaptive"** and **"learning"** in a good dictionary. Compare the uses of these words in different fields of control engineering.

8. **"Adaptive Control is not a mature technology."** Comment on this statement.

REFERENCES

(1) Astrom K. J., Wittenmark B., *Adaptive Control*, Pearson Education, Singapore, 2nd Edition, 1995.

(2) Katsuhiko Ogata, *Modern Control Engineering*, Prentice-Hall, Englewood Cliffs, New Jersey, 1970.

(3) I. J. Nagrath, M. Gopal, *Control Systems Engineering*, New Age International Publishers, New Delhi, 4th Edition, 2005.

Chapter 1

1.
$$e^{\mathbf{A}t} = \begin{bmatrix} \left(e^t - te^t + \frac{1}{2}t^2 e^t\right) & (te^t - t^2 e^t) & \left(\frac{1}{2}t^2 e^t\right) \\[2mm] \left(\frac{1}{2}t^2 e^t\right) & (e^t - te^t - t^2 e^t) & \left(te^t + \frac{1}{2}t^2 e^t\right) \\[2mm] \left(te^t + \frac{1}{2}t^2 e^t\right) & (-3te^t - t^2 e^t) & \left(e^t + 2te^t + \frac{1}{2}t^2 e^t\right) \end{bmatrix}$$

3. $y(t) = e^{-0.5t} \sin 0.5t$

5. $\mathbf{A} = \begin{bmatrix} 0 & 1 \\ -2 & -3 \end{bmatrix}; \ e^{\mathbf{A}t} = \begin{bmatrix} (2e^t - e^{-2t}) & (e^{-t} - e^{-2t}) \\ (-2e^{-t} + 2e^{-2t}) & (-e^{-t} + 2e^{-2t}) \end{bmatrix}$

7. $e^{\mathbf{A}t} = \begin{bmatrix} e^{\sigma t} \cos \omega t & e^{\sigma t} \sin \omega t \\ -e^{\sigma t} \sin \omega t & e^{\sigma t} \cos \omega t \end{bmatrix}$

9. $\begin{bmatrix} x_1 \\ x_2 \\ x_3 \end{bmatrix} = \begin{bmatrix} 0 & 1 & 0 \\ 0 & 0 & 1 \\ -4 & -4 & -3 \end{bmatrix} \begin{bmatrix} x_1 \\ x_2 \\ x_3 \end{bmatrix} + \begin{bmatrix} 0 & 0 & 0 \\ 0 & 0 & 0 \\ 1 & 3 & 4 \end{bmatrix} \begin{bmatrix} u_1 \\ u_2 \\ u_3 \end{bmatrix}$

$\begin{bmatrix} y_1 \\ y_2 \end{bmatrix} = \begin{bmatrix} 0 & 4 & 0 \\ 0 & 0 & 1 \end{bmatrix} \begin{bmatrix} x_1 \\ x_2 \\ x_3 \end{bmatrix} + \begin{bmatrix} 3 & 0 & 0 \\ 0 & 4 & 1 \end{bmatrix} \begin{bmatrix} u_1 \\ u_2 \\ u_3 \end{bmatrix}$

11. (a) Completely state controllable.

(b) Not completely state controllable.

13. Completely state controllable, completely output controllable; completely observable.

15. $\begin{bmatrix} x_1 \\ x_2 \end{bmatrix} = \begin{bmatrix} \dfrac{1}{6} + \dfrac{5}{2}e^{-2t} - \dfrac{5}{3}e^{-3t} \\ -5e^{-2t} + 5e^{-3t} \end{bmatrix}$

$y = 1 + 10e^{-2t} - 5e^{-3t}$

Chapter 2

1. (a) $F(z) = \dfrac{z(z+1)}{(z-1)^3}$ (b) $F(z) = \dfrac{z(z+a)}{(z-a)^3}$

(c) $F(z) = \dfrac{z \sinh \beta}{z^2 - 2z \cosh \beta + 1}$ (d) $F(z) = \dfrac{z(z - \cosh \beta)}{z^2 - 2z \cosh \beta + 1}$

3. (a) $F^*(s) = \dfrac{1}{1 - e^{-sT}}$ (b) $F^*(s) = \dfrac{1}{1 - e^{-aT}e^{-sT}}$

5. (a) $\dfrac{z \sin \omega T}{z^2 - 2z \cos \omega T + 1}$ (b) $\dfrac{z\,e^{-\alpha T} \sin \omega T}{z^2 - 2ze^{-\alpha T} \cos \omega T + e^{-2\alpha T}}$

(c) $\dfrac{z(z - \cos \omega T)}{z^2 - 2z \cos \omega T + 1}$ (d) $\dfrac{z^2 - ze^{-\alpha T} \cos \omega T}{z^2 - 2ze^{-\alpha T} \cos \omega T + e^{-2\alpha T}}$

7. $f(k) = \cos\left(\dfrac{k\pi}{2}\right); k = 0, 1, 2, \ldots$

9. $y(k) = 2\,(1)^k - 2\left(0.5\sqrt{2}\right)^k \cos \dfrac{\pi}{4}\, k$

11. $(a)\,\mathbf{A} = \begin{bmatrix} -2 & 0 \\ 0 & -3 \end{bmatrix}, \mathbf{B} = \begin{bmatrix} 1 \\ 1 \end{bmatrix}$ and, $\mathbf{C} = \begin{bmatrix} 1 & -1 \end{bmatrix}$

(b) STM $= \mathbf{A}^k = \begin{bmatrix} (-2)^k & 0 \\ 0 & (-3)^k \end{bmatrix}$ $(c)\, y(k) = \dfrac{1}{4}(-3)^k - \dfrac{1}{3}\,(-2)^k + \dfrac{1}{12}$

Chapter 3

1. Asymptotically stable at the origin.
3. Asymptotically stable in-the-large at the origin.
5. Asymptotically stable at the origin.

Chapter 4

3. $u^*(t) = -0.236\,\{x_1(t) + x_2(t)\}$

Appendix B

Laplace and z-Transform Pairs

Time Function $f(t)$	Laplace Transform $F(s)$	z-Transform $F(z)$
$\delta(t)$	1	1
$u(t)$	$\dfrac{1}{s}$	$\dfrac{z}{z-1}$
t	$\dfrac{1}{s^2}$	$\dfrac{Tz}{(z-1)^2}$
$\dfrac{t^2}{2}$	$\dfrac{1}{s^3}$	$\dfrac{T^2 z(z+1)}{2(z-1)^3}$
$e^{-\alpha t}$	$\dfrac{1}{s+\alpha}$	$\dfrac{z}{z-e^{-\alpha T}}$
$te^{-\alpha t}$	$\dfrac{1}{(s+\alpha)^2}$	$\dfrac{Tze^{-\alpha T}}{(z-e^{-\alpha T})^2}$
$1-e^{-\alpha t}$	$\dfrac{\alpha}{s(s+\alpha)}$	$\dfrac{(1-e^{-\alpha T})z}{(z-1)(z-e^{-\alpha T})}$
$\sin\omega t$	$\dfrac{\omega}{s^2+\omega^2}$	$\dfrac{z\sin\omega T}{z^2-2z\cos\omega T+1}$
$e^{-\alpha t}\sin\omega t$	$\dfrac{\omega}{(s+\alpha)^2+\omega^2}$	$\dfrac{ze^{-\alpha T}\sin\omega T}{z^2-2ze^{-\alpha T}\cos\omega T+e^{-2\alpha T}}$
$\cos\omega t$	$\dfrac{s}{s^2+\omega^2}$	$\dfrac{z(z-\cos\omega T)}{z^2-2z\cos\omega T+1}$
$e^{-\alpha t}\cos\omega t$	$\dfrac{s+\alpha}{(s+\alpha)^2+\omega^2}$	$\dfrac{z^2-ze^{-\alpha T}\cos\omega T}{z^2-2ze^{-\alpha T}\cos\omega T+e^{-2\alpha T}}$

C

ABOUT THE
ONLINE RESOURCES

- Included in the online resources are simulations and other files related to control systems topics
- See the "README" files for any specific information/ system requirements related to each file folder, but most files will run on Windows XP or higher
- The online resources are available for download at www.jblearning.com/catalog/9781934015216.

Index

www.ingramcontent.com/pod-product-compliance
Lightning Source LLC
Chambersburg PA
CBHW081809200326
41597CB00023B/4204